Studies in Big Data

Volume 116

Series Editor

Janusz Kacprzyk, Polish Academy of Sciences, Warsaw, Poland

The series "Studies in Big Data" (SBD) publishes new developments and advances in the various areas of Big Data- quickly and with a high quality. The intent is to cover the theory, research, development, and applications of Big Data, as embedded in the fields of engineering, computer science, physics, economics and life sciences. The books of the series refer to the analysis and understanding of large, complex, and/or distributed data sets generated from recent digital sources coming from sensors or other physical instruments as well as simulations, crowd sourcing, social networks or other internet transactions, such as emails or video click streams and other. The series contains monographs, lecture notes and edited volumes in Big Data spanning the areas of computational intelligence including neural networks, evolutionary computation, soft computing, fuzzy systems, as well as artificial intelligence, data mining, modern statistics and Operations research, as well as self-organizing systems. Of particular value to both the contributors and the readership are the short publication timeframe and the world-wide distribution, which enable both wide and rapid dissemination of research output.

The books of this series are reviewed in a single blind peer review process.

Indexed by SCOPUS, EI Compendex, SCIMAGO and zbMATH.

All books published in the series are submitted for consideration in Web of Science.

Nhien-An Le-Khac · Kim-Kwang Raymond Choo

A Practical
Hands-on Approach
to Database Forensics

Nhien-An Le-Khac ⓘD
School of Computer Science
University College Dublin
Dublin, Ireland

Kim-Kwang Raymond Choo ⓘD
Department of Information Systems
and Cyber Security
University of Texas at San Antonio
San Antonio, TX, USA

ISSN 2197-6503   ISSN 2197-6511   (electronic)
Studies in Big Data
ISBN 978-3-031-16129-2   ISBN 978-3-031-16127-8   (eBook)
https://doi.org/10.1007/978-3-031-16127-8

This Springer imprint is published by the registered company Springer Nature Switzerland AG
The registered company address is: Gewerbestrasse 11, 6330 Cham, Switzerland

Foreword

It is an honor to write a foreword on this book on database forensics by Kim-Kwang Raymond Choo and Nhien-An Le-Khac.

Database forensics in this book are well described, with many examples, and also the need for validation. Since the rate of change in digital evidence and also database forensics is fast, it is important to have the newest insights in the field in this book. I was excited to read this book with many viewpoints on database forensics in instant messaging, SQL databases, mobile phones and IoT devices. Case examples are important to explore and give insights for future cases in this field.

As a forensic scientist at the Netherlands Forensic Institute, as well as my chair Forensic Data Science at the University of Amsterdam, this is a valuable book for researchers, teaching and practitioners in the field. The book has covered different expert knowledge and can also be used as an excellent reference in these fields.

Zeno Geradts
Forensic Scientist, Netherlands Forensic Institute
Chair of Forensic Data Science
University of Amsterdam
Amsterdam, The Netherlands

Acknowledgements

This book would not have been possible for the amazing students enrolled in the Master of Science (M.Sc.) in Forensic Computing and Cybercrime Investigation program at the University College Dublin, Ireland—Katherine Moser, Jayme Winkelman, Benno Krause, Daniel Meier, Shuo Yan and Jacques Boucher, who were willing to dedicate their time and efforts to work on the research, share their findings and co-author chapters in this book.

We are also extremely grateful to Springer and their staff for their support in this project. They have been most accommodating of our schedule and helping to keep us on track.

We would like to thank our family and our loved ones for their unending support. To Thanh Thoa and Tri Nhien!

Contents

About the Authors

Nhien-An Le-Khac (IEEE Senior member) is Lecturer at the School of Computer Science, University College Dublin (UCD), Ireland. He is currently the Program Director of UCD MSc program in Forensic Computing and Cybercrime Investigation (FCCI)—an international program for law enforcement officers specializing in cybercrime investigations. He is also the co-founder of UCD-GNECB Postgraduate Certificate in fraud and e-crime investigation. Since 2008, he was Research Fellow in Citibank, Ireland. He obtained his Ph.D. in Computer Science in 2006 at the Institut National Polytechnique Grenoble, France. His research interest spans the area of Cybersecurity and Digital Forensics, Artificial Intelligence for Security, IoT Security, Fraud and Criminal Detection. He is an Associate Editor of Elsevier Internet of Things. He has published more than 150 scientific papers in peer-reviewed journal and conferences in related research fields, and his recent edited books have been listed the Best New Digital Forensics Book according to BookAuthority. He got a UCD Computer Science Outstanding Teaching Award in 2018.

Kim-Kwang Raymond Choo received the Ph.D. in Information Security in 2006 from Queensland University of Technology, Australia. He currently holds the Cloud Technology Endowed Professorship at The University of Texas at San Antonio. He is the founding co-Editor-in-Chief of ACM Distributed Ledger Technologies: Research and Practice, and the founding Chair of IEEE TEMS Technical Committee on Blockchain and Distributed Ledger Technologies, the Senior Editor of Forensic Science International: Digital Investigation, the Department Editor of IEEE Transactions on Engineering Management, and the Associate Editor of IEEE Transactions on Dependable and Secure Computing, and IEEE Transactions on Big Data. He is an ACM Distinguished Speaker and IEEE Computer Society Distinguished Visitor (2021–2023), and a Web of Science's Highly Cited Researcher (Computer Science—2021, Cross-Field—2020). In 2015, he and his team won the Digital Forensics Research Challenge organized by Germany's University of Erlangen-Nuremberg. He is the recipient of the IEEE Systems, Man, and Cybernetics Technical Committee on Homeland Security (TCHS) Research and Innovation Award in 2022, and the 2019 IEEE Technical Committee on Scalable Computing Award for Excellence in Scalable

Computing (Middle Career Researcher). He has also received best paper awards from IEEE Systems Journal in 2021, IEEE Computer Society's Bio-Inspired Computing STC Outstanding Paper Award for 2021, IEEE DSC 2021, IEEE Consumer Electronics Magazine for 2020, Journal of Network and Computer Applications for 2020, EURASIP Journal on Wireless Communications and Networking in 2019, IEEE TrustCom 2018, and ESORICS 2015, as well as the Outstanding Editor Award for 2021 from Future Generation Computer Systems.

Chapter 1
Databases in Digital Forensics

Nhien-An Le-Khac⑩ and Kim-Kwang Raymond Choo⑩

1.1 Introduction

While digital forensics (or referred to as cyber forensics in recent times) play an increasingly important role in our current society (or 'metaverse'), the role of databases in data/evidence acquisition cannot be understated [1–3], for example in mobile device/application forensics [4], Internet of Things (IoT) forensics [5], cryptocurrency forensics [6], etc.

There are, however, a number of operational challenges in identifying and acquiring data of forensic or evidential interest and relevance from the different databases on the devices and systems under investigation, as noted in several of the chapters in this book as well as our previous books [7, 8]. We observe that while there is only a small number of technical books on a narrow aspects of digital forensics, there are few edited or authored books on database security and forensic education. This is the gap we seek to address in this book.

1.2 Organization of This Book

In the next chapter, we will provide a high-level introduction of databases and their roles in digital forensics.

In Chap. 3, we focus on the examination of the Signal instant message application. We demonstrate what and how data can be acquired from the various databases installed on a laptop running Windows 10 Pro 64-bit, an iPad Air (4th Gen) running iOS 14.6, an iPhone 8 running iOS 14.6, an iPhone X running iOS 14.3, an Android Samsung Galaxy S20 Ultra, an Android Samsung Galaxy S10, using Magnet AXIOM, Oxygen Forensic Detective v13.6, Cellebrite UFED Touch 2 7.45.1, Cellebrite Physical Analyzer 7.46.

Similarly for Chaps. 4 and 5, we focus on the examination of the qTox messenger application and the PyBitmessage messenger application, and explain what and how data can be acquired from a range of devices and systems.

In Chaps. 6 and 7, we focus on the examination of an iPhone X (model A1901) running iOS 14.3 with an A11 chipset, an iPhone 5s (model A1533) running iOS 12.5.3 with an A7 chipset, an iPhone SE (model A1723) running iOS 14.6 with an A9 chipset, an iPhone 7 (model A1778) running iOS 10.2 with an A10 chipset, and an iPhone 7+ (model A1784) running iOS 11.0.3 with an A10 chipset, as well as several other IoT devices.

Finally, in the last chapter, we forensically examine Google Chrome browser in order to identify the types and range of forensic artifacts that could be recovered.

References

1. Chopade, R., & Pachghare, V. K. (2019). Ten years of critical review on database forensics research. *Digital Investigation, 29*, 180–197.
2. Choi, H., Lee, S., & Jeong, D. (2021). Forensic recovery of SQL server database: Practical approach. *IEEE Access, 9*, 14564–14575.
3. Al-Dhaqm, A., Ikuesan, R. A., Kebande, V. R., Razak, S., Grispos, G., Choo, K.-K. R., Al-Rimy, B. A. S., & Alsewarim, A. A. (2021). Digital forensics subdomains: The state of the art and future directions. *IEEE Access*, 152476–152502.
4. Fukami, A., Stoykova, R., & Geradts, Z. (2021). A new model for forensic data extraction from encrypted mobile devices. *Forensic Science International: Digital Investigation, 38*, 301169.
5. Sandvik, J.-P., Franke, K., Abie, H., & Årnes, A. (2022). Quantifying data volatility for IoT forensics with examples from Contiki OS. *Forensic Science International: Digital Investigation, 40*, 301343.
6. Koerhuis, W., Kechadi, T., & Le-Khac, N.-A. (2020). Forensic analysis of privacy-oriented cryptocurrencies. *Forensic Science International: Digital Investigation, 33*, 200891.
7. Zhang, X., & Choo, K.-K. R. (2020). *Digital forensic education. An experiential learning approach*. Springer.
8. Le-Khac, N.-A., & Choo, K.-K. R. (2020). *Cyber and digital forensic investigations*. Springer International Publishing. https://doi.org/10.1007/978-3-030-47131-6

Chapter 2
Database Forensics

Nhien-An Le-Khac⬥ and Kim-Kwang Raymond Choo⬥

2.1 Introduction to Databases

Today, investigators need databases to store and analyze forensic and criminal data (Fig. 2.1). Hence, they should design and build database solutions for investigations.

Besides, most parts of digital forensics today deal with extraction and collection of evidences from databases such as history or cookies information of browsers [1], account information, contact list or call logs of VoIP (Voice over Internet Protocol) application (Fig. 2.2) [2, 3] and social media apps [4, 5]. Therefore with a best understanding of database structure, investigator could retrieve evidences more efficient from variant types of data across electronic devices.

2.1.1 What Is a Database?

Database is not only tables as shown in previous examples. To define what a database is, we should know what is data? So, Eliot said: "Where is the wisdom? Lost in the knowledge. Where is the knowledge? Lost in the information" [6], and Mr. Celko, he said: "Where is the information? Lost in the data" [7]. Maybe the poet Eliot never wrote a computer program in his life but Mr. Celko did. However, we agree with both Eliot and Celko on their points about unofficial definition of knowledge, information and data. Actually, data is a representation of fact, figure or idea. Data normally refers to a collection of numbers, characters, image, etc.

CHICAGO CRIMES

Case Number	date	Block	Primary Typ	Description	Location De	Arrest	FBI Code
HL290758	07/02/2011	057XX S BISHOP ST	HOMICIDE	FIRST DEGREE MURDER	PORCH	Yes	01A
HR657368	13/01/2011	010XX W 79TH ST	CRIMINAL TRE!	TO LAND	OTHER	No	26
HS103084	01/03/2011	030XX W LEXINGTOI	THEFT	FROM BUILDING	ABANDONED E	No	6
HS232857	04/03/2011	063XX S ARTESIAN A	NARCOTICS	POSS: CANNABIS 30GM:	SIDEWALK	Yes	18
HS239712	21/01/2011	082XX S ELLIS AVE	BATTERY	DOMESTIC BATTERY SIM	APARTMENT	No	08B

Investigations

CaseNo	CaseType	OIC
3692	Grand Theft Auto	284
3693	Murder	132
3694	Blackmail	284

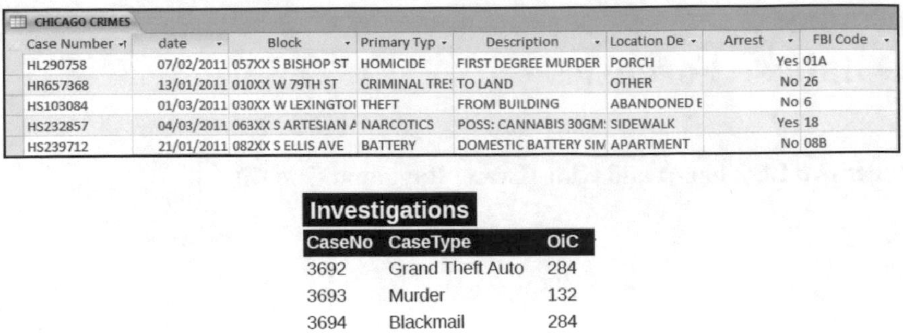

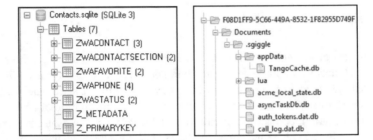

Fig. 2.1 Examples of forensic and criminal databases

Fig. 2.2 VoIP apps databases

Besides, when data has a relational connection, its meaning is an information. For example, Bob, 10 are data. "Bob" is "10" years old is an information. Knowledge is the appropriate collection of information such that it is to be useful. For example, when we have a collection of information such as "Bob is 10 years old", "Alice is 5 years old", "Rian is 6 years old", etc. we can have a rule "80% of children in this room is less than 8 years old". This rule is a knowledge.

Data is typically processed by human or stored and processed in computer as files. But how these files are organized so that users can access their information easily and effectively at any given time? It could be a database.

There are many way is define a database. Firstly, it is a collection of data, which is structurally stored in a computer system. A database can be considered a collection of related data which are describing the activities of relevant organizations. For instance, a school database contains information about the pupils, school, subjects, and rooms. This database also has the relationships between these objects such as school teaching subjects and the use of rooms for subjects. At the technical point of view, a database could be a tool that stores data, and allows users to create, read, update, and delete the relevant data per request [7].

Databases can be simple or very complicated [8]. We can have not only Universities' databases as discussed previously but also many examples of databases in everyday uses. For example, a telephone company has a customer database with basic information about their clients such as first name, surname, address, city, postal code and phone number.

An example of large databases is databases existing in banking systems, library catalogues, hotel or airline reservation, social networking, etc. Besides, databases can also be used to store images, audio streams and videos digitally.

Law enforcement has been used databases to store and manage criminal data, which can help to identify and catch. Also, by suing the databases, law enforcement members can effectively gather relevant information to assist in investigations.

2.1.2   Database Management System (DBMS)

A Database Management System (DBMS) [9] is a software system specifically designed to handle and exploit databases at different scales. Most DBMSs are used to manage relational databases. But why we need a DBMS? Today, we live in a world experiencing information explosion. In order to manage efficiently the huge amount of data, we need DBMSs because a DBMS can provide:

- Data independence: A DBMS gives an abstract view of data representation and storage to the application programs.
- Efficient data access: A DBMS normally applies optimal techniques to handle and access data efficiently.
- Data integrity and security: A DBMS implements a variety of mechanisms to guarantee the integrity constraints on the data and to control the access the relevant data.
- Data administration: A DBMS allows the databases to share among several users or different user groups in an efficient way in terms of storing and retrieving.

A DBMS moreover can handles concurrent access and crash recovery as well as to reduce application development time. However, a DBMS has some drawbacks such as its overhead costs.

Some popular DBMSs in the market can be listed as Microsoft SQL Server [10], Oracle database (Oracle DB) [11], PostgreSQL [12] (open source) and MySQL [13] (open source).

There are four important elements of a DBMS: modeling language, data structures, data query language and transactions. A data model is a collection of data description at a high-level that hides details in low-level storage. A data model can be represented by data schemas. A modeling language is used to define the schema of each database stored in a DBMS, according to the data model. As mentioned previously, most DBMSs today are using the relational data model. There are moreover other data model such as: entity-relationship model, relational algebra, hierarchical model, network model and object data model.

Data structure relates to data types, relationships and constraints and the data structures allows DBMS to interact with the data align to their integrity.

A query language is a specialized language that allows to post a queries. A query is a question involving the data stored in a DBMS. For example, some queries can be posted for the University database such as "What is the student name of the student with an ID 1234?", "How many students are enrolled in course COMP47370?" etc.

A transaction is an execution from the user program in a DBMS. It is also a basic unit of change in the DBMS.

2.1.3 Database Types and Users

Let's have a look at different types of a database [14]. First, it is a flat file. A flat file is simply file containing text. A flat file could be a database if we add structure in. For example, by separating values with commas, a flat file becomes a .csv file. Although flat files do not provide many services, they are simple and easy to understand. Flat files are also good places to store configuration settings such as .INI files.

In a hierarchical database, records' relationships form a tree-like structure. One data record logically links to other data records. The structure of hierarchical database is simple and it is restricted to a one-to-many relationship. An example of a hierarchical database is the Windows system registry (Fig. 2.3).

A network database is not a database that is used over the network of computers. It relates to the database structure that consists of a collection of *nodes* connected by *links*. The nodes and links represent objects such as members of a social network. This database structure is uncommon and normally used to express many-to-many relationships [14].

The object oriented database is used to manage objects of varied types such as pictures, video clips, voice and text, etc. The object database management systems (ODBMS) provides special query syntax for accessing and retrieving objects from the database. This database type is popular for Web-based applications.

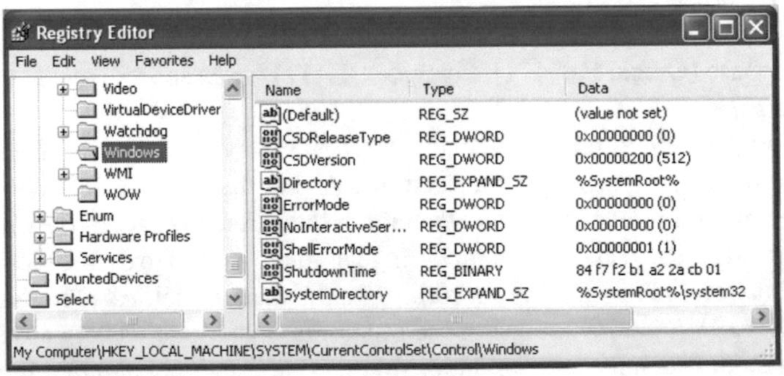

Fig. 2.3 Example of windows registry

Regarding the database users, there are some popular roles in practice such as a database administrators (DBA), database designers, System Analysts and Application programmers and finally end users.

2.2 Relational Databases

2.2.1 Basic Concepts

In the relational data model, data is organized in tables [8, 9, 14]. A table has a set of records (or rows) and each record can have many attributes/fields (or columns). The simple structure of relational data model makes it easy to understand and easy to exploit with high-level languages such as Structured Query Language (SQL) [15] to query the data. Current popular relational DBMSs include Oracle DB, MS SQL Server, MySQL, DBs, etc.

Figure 2.4 illustrates an example of a relational database, which is a university database to explain the following concepts: *table*, *row* and *column*. Within this database will be three tables named *Student*, *Module* and *Enroll*. Table *Student* for example includes student information such as student id (SID), name, age and grade point average (*gpa*). Each information is described as a field (or an attribute, or a column) of this table. A column is also called an attribute. The set of the validate values of an attribute is called the attribute's domain. It relates to data type of this field. For example, the domain of the attribute age is integer, the domain of the attribute *name* is string and of the attribute *gpa* is real number. Besides, a table is sometime called a relation.

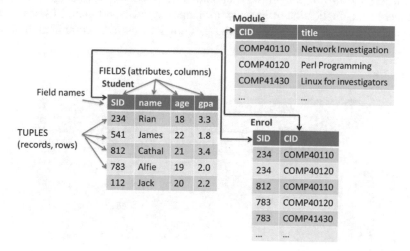

Fig. 2.4 Example of tables in a relational database

Schema versus Instance

At the model level, there are two concepts related to relational database: schema and instance.

A schema describes the column head for the table that specifies how data to be structured at the logic level. The schema is also called meta-data. In fact, a schema species the name of table, all attribute (field) names, and the domain for each attribute. The schema is normally defined at the setup time of a relational database and it's rarely changed because of high cost for updating of instances.

For example, for the Student table in previous example, we have the Student schema as you can see here, the name of this schema is also Student and this schema specifies the name and data type of each attribute.

A record/row in a table is called a table instance, it is also called a tuple. In fact, all tuples in a table have the same number of fields/attributes and there is no two tuples that are identical. That means a table contains a set of unique instances. Besides the order of these instances are not important. The value of an instance can be updated but it always conforms to its schema. In table *Student* for example (Fig. 2.4), there are five instance such as (234, Rian, 18, 2.3), (541, James, 22, 1.8), etc.

Keys and Index

Key is an important concept of relational database. A key is a combination of one or more fields that it can be used to identify records (rows) in a table. A key of a table is defined as a set of one or more attributes (or columns) in this table. For example, a *Student* table can use SID as a key to find students. If a student's ID is known, the relevant student's record can be found in the table. Hence, this table, the key is SID.

A compound/composite key is a key that has more than one attributes (or columns). For example, in table *Case_Investigation* (Fig. 2.5), it might be used the combination of **CaseID** and **Investigator** as a compound key. Of course there is an assumption that there is no duplication of investigator names in the department. Besides, in the previous example of University's database, it might be used the combination of SID and CID to look up enrollments.

Case_Investigation

Case ID	CriminalType	Description	Location	Investigator	Date
1	Criminal damage	to property	Residence	Jack Bloggs	2030-10-11
1	Criminal damage	to property	Residence	Anne Smith	2030-10-11
2	Burglary	forcible entry	Apartment	James Doha	2031-09-10
2	Burglary	forcible	Apartment	Jack Bloggs	2031-09-10
3	Theft	from building	Bar or tavern	John Brown	2020-05-15

Fig. 2.5 Case_Investigation table

A *superkey* is a set of one or more attributes in a table so that a record is unique. There are no two records that can have the same values. Hence, a *superkey* is also called a unique key. For example, in the *Enroll* table (Fig. 2.4), the SID and CID attributes together form a *superkey* because no two records have exactly the same SID and CID values. Besides, in the *Student* table, the *superkey* is SID.

There are more key definition in relational database. A *primary key* is a superkey and a table can have only one primary key. It should be noted that every records in a database has the own primary key.

Besides, a secondary key is used to lookup records but it does not guarantee the record uniqueness.

Another kind of key is the foreign key, which is used to refer to a primary key of another table.

Index is another important concept of the relational databases. This special database structure allows to find records quicker and easier by using one or more attributes' values. Note that indexes are not the same as keys. For example, as shown in Fig. 2.4, *Student* table holds student information: name, age, and gpa. This table also has the primary key SID. If students do not remember their student IDs, the student name can be used to search a student. If this table is indexed by name, users can quickly locate a student's record in two ways: by student ID or by name.

2.2.2 Database Design

Databases store vast quantities of information. Consider government social security databases for instance. Information is stored on every citizen in the entire nation. This information includes name, address, date of birth, income, tax status et cetera. Searching through all of this information can be very time consuming. If every record needed to be checked in the database it would be a very inefficient system. As mentioned above, there are two ways to make these searches faster: using keys or indexes. For example, the social security system gives each person in the nation a unique identifying number: a social security number. This number acts as a key. It is a unique number that identifies the person in question and can be searched for very quickly in the database. Rather than search for the name for example, Jim Murphy in the database users can search for the unique social security number.

There are three main steps in database design [9]. Gathering the required information, in other words to make sure that all of the necessary information can be stored in the database. These pieces of information will form the fields. This information should also be logically divided into tables. For each of the fields identified designers should select a datatype for this. Finally designers need to create keys or indexes to make retrieval quicker.

2.3 Structured Query Language (SQL)

2.3.1 SQL and SQLite

SQL, or Structured English Query Language [15] is the standard query language for relational DBMSs. There are different versions of SQL: SQL-86 or SQL-1 is the first standard version of SQL. SQL-2 is a revised and expanded version of SQL-86, it is also called SQL-92. The next version is a well-recognized standard SQL-99. The other standards such as SQL-3 have been proposed. However they are not fully endorsed by the industry.

SQL includes statements for data definitions, queries and updates. SQL uses the term table, row and column for the relation model.

For example, the CREATE command in SQL is for the data definition that is used to create schemas, tables, relations, domains, etc. The ALTER command is used to change the definition of a table such as adding or dropping a column, changing a column definition (name of column, data type), and adding or dropping table constraints. The DROP command is used to drop tables, domains or constraints. The SELECT command in SQL is a basic statement for retrieving information from database.

SQLite is public-domain, lightweight relational DBMS [16]. SQLite follows nearly entire SQL-92 standard. SQLite is designed by D. Richard Hipp and the first version was released in August 2000.

SQLite does not require a separate server as other DBMSs such as MySQL, SQOL server of Oracle to operate. That means users do not need to setup/configure a server. It is stability, ease to use as for instance, creating a SQLite database is as simple as opening a file. Moreover, the whole database are stored in a single file that can run on many platforms. SQLite can runs with a limit of hardware resource (CPU, RAM).

Recently, many applications are now storing log information in SQLite 3 database format. These include: Skype; WhatsApp; iOS; Google-Chrome; Mozilla Firefox and many more. So, the ability to extract this log information and handle it in meaningful ways is essential in forensics.

2.3.2 SQLite Basic Commands

The command line interface for SQLite available for all major. There are also some SQLite management commands that start with a dot as follows:

.tables: display all tables in the database
.schema: display the schema of tables created
.quit to exit the SQLite application

There are also some basic SQLite commands as follows[1]:

- Launch SQLite: sqlite3
- Create/open a SQLite file: sqlite3 <filename>
- Create a table: CREATE TABLE command
- Add a record into a table: INSERT INTO command
- Delete a table: DROP TABLE command
- Retrieve information: SELECT command
- Delete a record: DELETE command

Examples:

- Create/open university database:

  ```
  sqlite3 university
  ```

- Create Student table:

  ```
  CREATE TABLE Student (SID   INTEGER,
                        name  TEXT,
                        age   INTEGER,
                        gpa   REAL);
  ```

- Add a record into Student table:

  ```
  INSERT INTO Student VALUES (1, "Rian", 18, 3.3);
  ```

- Retrieve all students who have *gpa* greater than 3.68:

  ```
  SELECT * FROM Student WHERE gpa > 3.68;
  ```

- Delete student James:

  ```
  DELETE FROM Student WHERE name = 'James';
  ```

- Delete Student table:

  ```
  DROP TABLE Student.
  ```

2.4   Database Forensics

As mentioned in Sect. 2.1, there are many applications varied from web browsers to mobile apps, running on different operating systems and platforms that are using databases, especially SQLite databases to store relevant information such as searching history or cookies information of browsers, account information or call logs, which are important artifacts for any forensic investigation. This also means

---

[1] More details of SQLite commands' syntax and examples can be found in: https://www.sqlite.org/docs.html.

that investigators regularly encounter such artifacts from databases. Hence, how to conduct the database forensics from a given application? How to acquire and analysis artifacts?

Figure 2.6 illustrates a flow chart of the suggested forensic method that has three main phases: preparation, acquisition and analysis.

The investigator should explore the application in the first phase (Step 1, Fig. 2.6). The objective is to understand the application i.e. what can the application do? What information might be stored in its databases (e.g. SQLite files)? To achieve this objective, investigators should look for information related to the application, using application websites, open sources, published documents, forensic sites and blogs etc. Without knowledge of what the application does it is very difficult to know what to look for. Hence, in some cases, using the application (Step 1a, Fig. 2.6) is recommended to gain a better understanding of this application.

The second phase includes three steps: identify the database files, collect database information and extract relevant information from databases.

The objective of identifying the database files (Step 2, Fig. 2.6) is to locate all databases used by the application. Investigators can look at open sources on the internet. Databases of existing applications have probably been illustrated in published documents, relevant websites, blogs, etc. Moreover, Profiles, Preferences folders in applications could store location information. Another approach is to install the application in a virtual machine or a simulated platform. Investigators should take snapshots of (database) files existing before and after the installation and try to identify the new (database) files in the after snapshot. The outcome of this step is a list of all database files used by this application.

Next, investigators should collect as much as much information on the databases located (Step 3, Fig. 2.6). The relevant information of each database can be listed as a list of tables, the structure of all tables, the number of entries in all tables, the relationship between tables, index and key for each table, etc. To assist the collection information in this step, a good practice is to use the application by performing its common tasks (Step 3a, Fig. 2.6) then repeat Step 3 to identify which information created or updated, which table entries have been modified and examine these changes.

The final step of this acquisition stage is to launch SQL queries to extract necessary data (artifacts) from all relevant tables (Step 4, Fig. 2.6) of the databases. Note that the storage of the acquired data and devices should follow the policy regarding the chain of custody and jurisdiction.

The acquired data are analyzed in Stage 3 (Step 5, Fig. 2.6). Note that investigators could re-run Step 4 (Fig. 2.6) if the acquired data outcome is un-sufficient for the investigation.

Finally, when the analysis phase is finished, investigators should conduct a review of the process and the actions of the previous steps to validate the investigation process. A formal report is produced in this phase (Step 6, Fig. 2.6) to record all the steps of the investigation, explain the findings etc.

Fig. 2.6 Database forensic
flowchart

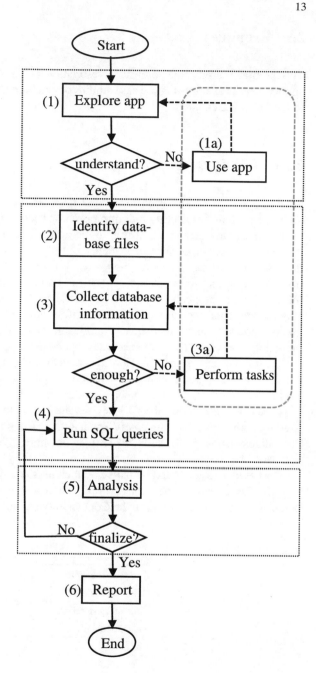

2.5   Examples

This section presents the forensic investigation of databases from some popular applications as examples, more case studies of database forensics will be introduced in the following chapters.

2.5.1   IOS Database Investigation

Smartphones are widely used today and there are security risks associated with their use such as conducting a digital crime or becoming a victim of one. Hence, smartphones have an important role in the crime investigation [17]. Eventually, in crime scenes the importance of evidence out of the smartphone is increasing with the effect that more and more phones are seized and for a thoroughly investigation offered at the digital department of the police. In the early days the possible information out of a cell phone was phone calls, SMS and contacts. Nowadays the smartphone contains email, web browsing information, Chat, Instant Message, Documents, Photos, Videos, and plenty of applications used daily [18]. The most popular operating systems on smartphones today can be listed as iOS and Android.

iOS (or iPhone OS up to Version 3) is developed by Apple specifically for many Apple's products such as iPhone, iPad, iPod touch, Apple TV, etc. iOS revolutionized the way cell phones have been created. Like on other mobile platforms, most important information on the iOS devices are stored in SQLite databases that are used by both native and third-party applications. The popular databases in iOS are the Address Book, SMS, and Call History databases. Table 2.1 lists common iOS databases in different iOS versions.

Most evidence generated by native applications are located in the *Library* directory. In this directory, the *AddressBook* refers to the information related to the personal contacts that present in the Contact application. There are two databases of

Table 2.1 List of common iOS databases	SQLite databases	iOS12	iOS13	iOS14
	Addressbook.sqlite	√	√	√
	Calendar.sqlitedb	√	√	√
	Callhistory.storedata	√	√	√
	consolidated.db	√	√	√
	sms.db	√	√	√
	notes.sqlite	√	√	√
	photos.sqlite	√	√	√
	voicemail.db	√	√	√
	healthdb.sqlite	√	√	√
	tcc.db	√	√	√

interest in this *AddressBook* directory: *AddressBook.sqlitedb* and *AddressBookImages.sqlitedb*. The *AddressBook.sqlitedb* contains the information for each contact such as name, surname, phone number, e-mail address, etc. The number of tables in this database depends on iOS version. On the left of Fig. 2.7 is a list of tables from the contact database *AddressBook.sqlitedb*, which can be found in the *Home/Library/AddressBook* folder of recent iOS versions. The tables of interest are mainly *ABPerson* and *ABMultiValue*. *ABPerson* table (Fig. 2.7, right) with 46 fields contains the name, organization, department, and other general information for each contact. *ABMultivalue* table contains phone numbers, email addresses, website URLs, and other data for the case a contact may have more than one.

Some multivalued entries contain multiple values. For example, an address consists of a city, state, zip code, and country code and these values can be found in *ABMultiValueEntry* table. This table has *parend_id* field, which corresponds to the *rowid* of *ABMultiValue* table.

Figure 2.8 illustrates an example of retrieving some important information from tables of *AddressBook.sqlitedb* database.

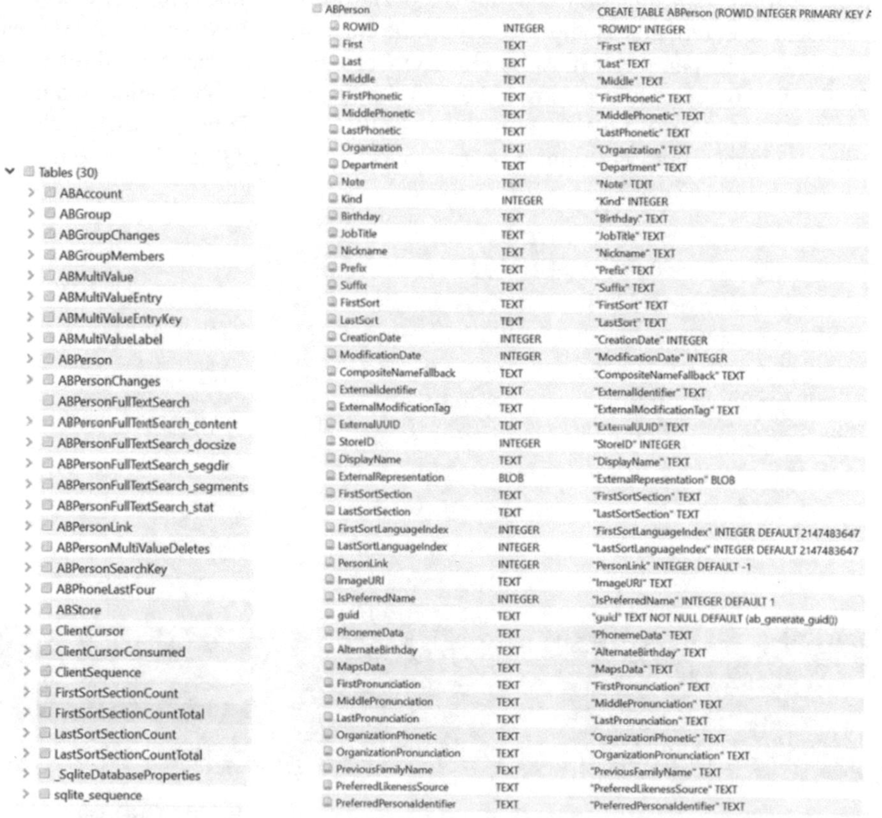

Fig. 2.7   *AddressBook.sqlitedb* tables (left) and list of fields of *ABPerson* table (right)

```
sqlite> SELECT Last,First, Middle, JobTitle, Department, Organization,
   ...>        Birthday, CreationDate, ModificationDate, ABMultiValueLable.value,
   ...>        ABMultiValueEntry.value, ABMultiValue.value
   ...> FROM   ABPerson, ABMultiValue, ABMultiValueEntry, ABMultiValueLabel
   ...> WHERE  ABMultiValue.record_id = ABPerson.rowid AND
   ...>        ABMultiValueLabel.rowid = ABMultiValue.label AND
   ...>        ABMultiValueEntry.parent_id = ABMultiValue.rowid;
```

Fig. 2.8 Exploring a suspect's contact information with a SQLite query

AddressBookImages.sqlitedb contains images associated to a given contact. In this database, the important table is *ABFullSizeImage*.

The Call History information is stored in the Call History database: *callhistory.storedata* that contains each of the missed, placed, and received calls, etc. on the device. So, this information helps to find tracks about incoming, outgoing and missed calls with time and date occurred and their duration. Tables contained within this database are listed in Fig. 2.9 (left) where ZCALLRECORD (right) is the important one.

Figure 2.9 (bottom) shows an example of exploring some important fields of ZCALLRECORD table such as the call sequence id, phone number, the duration of the call, the call direction, the call status and the call timestamp. The headers command is used in this example to display the column headers in our query results. The value of its parameter must be either ON or OFF. Note that the call direction showed in this example is a number i.e. 0, 1. To display this information in a more comprehensive format for the query result, ZORIGINATED field is used with the value 0 is "Incoming" and value 1 is "Outgoing" with the SQLite CASE command

Fig. 2.9 *callhistory.storedata* tables (left) and list of fields of ZCALLRECORD table (right). SQLite query to explore ZCALLRECORD table (bottom)

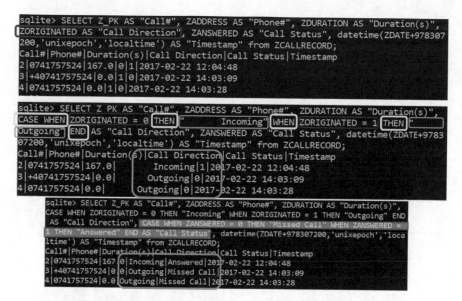

Fig. 2.10 More examples of SQLite queries to explore ZCALLRECORD table

(Fig. 2.10). Similarly, in the last example, the CASE command for the call status ZANSWERED field is used, with value 0—a missed call and 1—a call answered (Fig. 2.10).

Messages are one of the most significant data items to be recovered from iOS devices. The database, which is used to storedsuch information in iOs is *sms.db* that can be found in the *Home/Library/SMS* folder and the important table is *message* table (Fig. 2.11). This table contains the message, date and time, and whether the message was sent or received, etc. For example, this table uses specific fields for each status such as *is_sent*, *is_read*, *is_delievered*, etc. and the value of each specific field is either 0 (default) or 1 (Fig. 2.11).

From the forensic point of views, there are also other interesting sqlite databases in iOS such as *notes*, *photos*, *voicemail*, *healthdb* and *tcc.db* (Table 2.1). *tcc.db* database for example, tcc means Transparency Consent and Control system, it contains all the prevailing settings for privacy controls, including the allow lists which are displayed in the Privacy tab of the Security and Privacy pane. The *healthdb.sqlite* database contains all information collected or received by the Health app. The *voicemail.db* database contains voicemail entries. The *photos.sqlite* database stores a lot of photo asset information including location, date, time, etc.

Note database *NoteStore.sqlite* is also interesting because some users tend to store passwords and other important info in notes. Figure 2.12 illustrates tables in the *Note-Store.sqlite* database, and the important tables are ZNOTE and ZNOTEBODY. Some important fields are ZAUTHOR that contains the author's email address, ZTITLE contains the title of the note and ZCONTENT field of the ZNOTEBODY table has the content of the note. However, most of the notes' contains in the recent iOS version

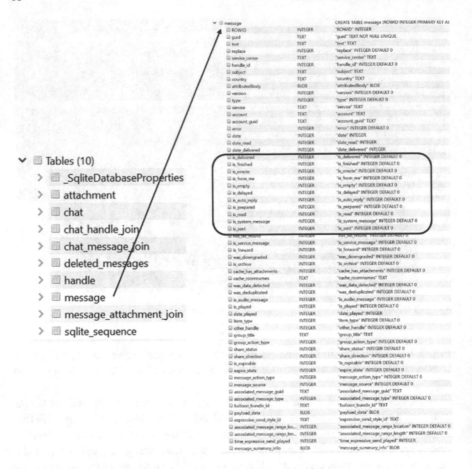

Fig. 2.11 *sms.db* tables (left) and list of fields of *message* table (right)

can be found in *NoteStore.sqlite* database that can be synchronized with the iCloud account and the note content is normally encrypted.

Figure 2.13 lists tables in the *Photos.sqlite* database, and an important table is ZGENERICASSET. There are many fields in this ZGENERICASSET table. Some important fields are ZDATECREATED, which is the timestamp of the image, ZLATITUDE and ZLONGITUDE, which are the location where the photo is taken, ZFILENAME and so on. Figure 2.13 also has an example of SQLite query to show all photo names with the timestamp and location.

Finally, Fig. 2.14 shows tables of *voicemail.db* database with an example of a SQLite query to get a list of voicemails sorted by date.

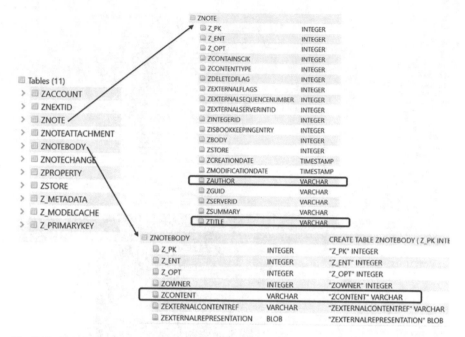

Fig. 2.12 *NoteStore.sqlite* tables (left) and list of fields of ZNOTE table (right) and ZNOTEBODY table (bottom)

2.5.2   WhatsApp Database Forensics

WhatsApp is an Instant Messaging and Voice over IP (VoIP) app developed by Brian Acton and Jan Koum. Users can use WhatsApp to exchange instant messages, images, video and audio media messages. Today, there are around two billion WhatsApp registered users active monthly. Hence it becomes an important source of forensic investigation [19].

WhatsApp is available on different devices such as iPhone, iPad, Android phones and tablets and also with different platforms such as Windows, Mac, iOS, Android, etc.

On iOS systems, the manual file system analysis initially shown that the WhatsApp files are in the directory *group.net.whatsapp.WhatsApp.shared/*. The activity and contact information are stored in SQLite database files. For example, the communication activity is stored in *ChatStorage.sqlite* database and the contact information is stored in the *ContactsV2.sqlite* database. There are other artefacts in *net.whatsapp.WhatsApp/Documents/* folder. The *ChatStorage.sqlite* database (Fig. 2.15) might be of interest in an investigation. Its most significant tables are ZWACHATSESSION, ZWAGROUPINFO, ZWAGROUPMEMBER, ZWAMEDI-AITEM and ZWAMESSAGE. ZWACHATSESSION table contains a list of unique conversations started with different contacts or groups. ZWAGROUPINFO table stores a list of group conversations. ZWAGROUPMEMBER table has a list of

Fig. 2.13 *Photos.sqlite* tables (left) and list of fields of ZGENERICASSET table (top right). SQLite query to show all photo names with the timestamp and location (bottom right)

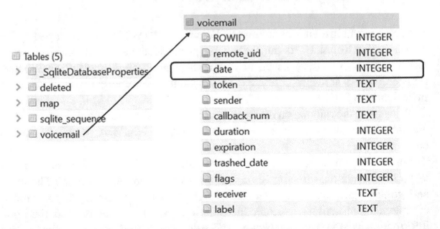

Fig. 2.14 Tables of *voicemail.db* (left), list of fields of *voicemail* table (right). SQLite query to explore *voicemail* table (bottom)

Fig. 2.15 List of tables in
ChatStorage.sqlite database

Tables (17)

> ZWABLACKLISTITEM
> ZWACHATPROPERTIES
> ZWACHATPUSHCONFIG
> ZWACHATSESSION
> ZWAGROUPINFO
> ZWAGROUPMEMBER
> ZWAGROUPMEMBERSCHANGE
> ZWAMEDIAITEM
> ZWAMESSAGE
> ZWAMESSAGEDATAITEM
> ZWAMESSAGEINFO
> ZWAPROFILEPICTUREITEM
> ZWAPROFILEPUSHNAME
> ZWAVCARDMENTION
> Z_METADATA
> Z_MODELCACHE
> Z_PRIMARYKEY

contacts participating in group conversations. ZWAMEDIAITEM table contains a list of exchanged media items and finally ZWAMESSAGE is an important one, it stores a list of messages exchanged.

Figure 2.16 illustrates more details of ZWAMESSAGE table. The interesting fields are: ZMESSAGESTATUS contains the message status. For example, value '1' is 'received', value '8' is 'read'. ZMESSAGETYPE is the message type, with '0' is text, '1' is image; '2' is video; '3' is audio; '4' is contact; '5' is location; '6' is group; '7' is URL; '8' is file; etc. ZGROUPMEMBER stores a value, which is a primary key in ZWAGROUPMEMBER table, it links with ZWAGROUPMEMBER to identify the sender in a group. ZMESSAGEDATE contains the created date or the received date of a message. ZSENTDATE contains the sent date of a message. And finally ZTEXT contains text or Emojis.

Another interesting WhatsApp database is *Contacts.sqlite* with the most important table is ZWAADDRESSBOOKCONTACT (Fig. 2.17). In this table, we can find the list of all contacts found on the phone with the phone numbers and the contact's profile. Some important fields are ZFULLNAME stores the Contact's full name. ZPHONENUMBER stores the contact's phone number. ZWHATSAPPID stores the contact's number in WhatsApp id format, i.e. the phone number in international format.

On the Android systems, WhatsApp artifacts can be found in the following folders: *data/com.whatsapp/databases/msgstore.db; data/com.whatsapp/databases/wa.db; data/com.whatsapp/...*

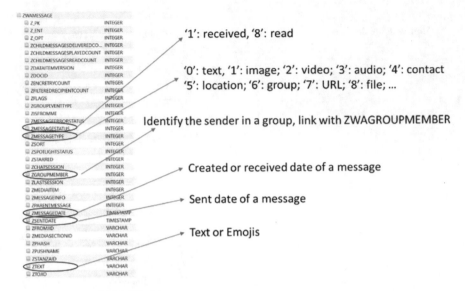

Fig. 2.16 ZWAMESSAGE table

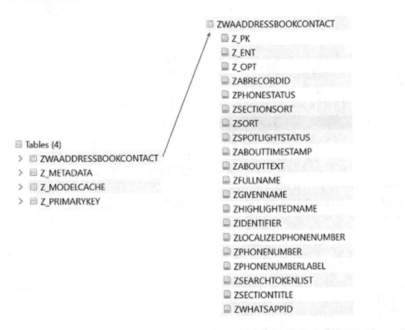

Fig. 2.17 *Contacts.sqlite* tables (left) and fields of ZWAADDRESSBOOKCONTACT table

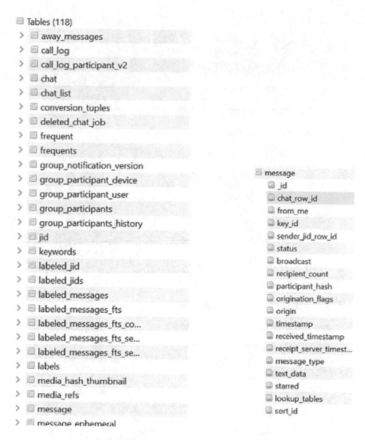

Fig. 2.18 *msgstore.db* tables (left) and fields of *message* table (right)

The folder 'databases' was identified as storing the most valuable information in order to reconstruct the communication history in WhatsApp, which were the two main database files *wa.db* and *msgstore.db* that contained all the exchanged attachments, such as images, video and contact cards. Figure 2.18 illustrates the structure of database and fields of its *message* table.

The *wa.db* database has one important table *wacontacts* (Fig. 2.19). This table contains a list of all contacts found on the phone (not only WhatsApp contacts). It has the following attributes:

- jid: The contact or group WhatsApp id in < phonenumber> @s.whatsapp.net format
- is_whatsapp_user: Boolean if the phone number is a registered WhatsApp user
- status: The status text of the contact
- number: The contact's phone number
- display_name: The contact's display name in the contact list
- phone_type: Number mapped to phone type, i.e. mobile, home etc.

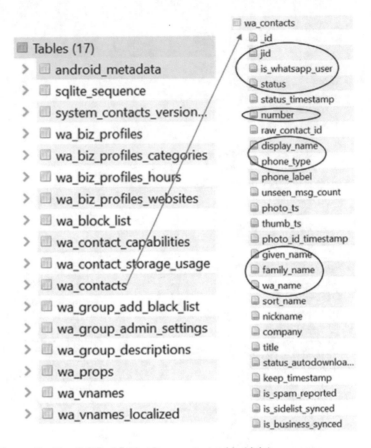

Fig. 2.19 *wa.db* tables (left) and fields of *wa_contacts* table (right)

- given_name: Contact's first name
- family_name: Contact's family name
- wa_name: Contact's WhatsApp name.

We refer readers to other related research efforts on WhatsApp forensics, such as those outlined in [19, 20].

2.6 Summary

This chapter present background of databases and database investigation that are necessary to follow case studies in following chapters. Basic concepts of databases including relational databases, database design and SQL language were introduced. This chapter also described a process for investigating application databases. Finally, two examples of database forensics were illustrated to show how to examine databases from iOS and from an instant message application. The following chapters will present case studies where database investigation techniques are applied in acquire relevant artifacts from different applications of forensic cases.

Acknowledgements Authors would like to thank Ranul Thantilage, who helped us to prepare relevant SQLite databases used in this chapter.

References

1. Warren, C., El-Sheikh, E., & Le-Khac, N.-A. (2017). Privacy preserving internet browsers—forensic analysis of browzar. In K. Daimi, et al. (Eds.), *Computer and network security essentials* (pp. 369–388 (18 pages)). Springer Berlin Heidelberg. https://doi.org/10.1007/978-3-319-58424-9_21

2. Sgaras, C., Tahar Kechadi, M., & Le-Khac, N.-A. (2015). *Forensics acquisition and analysis of instant messaging and VoIP applications. Lecture Notes in Computer Science* (Vol. 8915, pp. 188–199 (12 pages)). https://doi.org/10.1007/978-3-319-20125-2_16

3. Schipper, G. C., Seelt, R., & Le-Khac, N.-A. (2021). Forensic analysis of matrix protocol and Riot.im application. *Forensic Science International: Digital Investigation, 36*, 301118. https://doi.org/10.1016/j.fsidi.2021.301118

4. Thantilage, R., & Le-Khac, N.-A. (2020). Forensic analysis of e-dating applications based on iPhone backups. In G. Peterson & S. Shenoi (Eds.), *Advances in digital forensics XVI. Digital forensics 2020. IFIP advances in information and communication technology* (Vol. 589). Springer. https://doi.org/10.1007/978-3-030-56223-6_12

5. Voorst, R. V., Kechadi, T., & Le-Khac, N.-A. (2015). Forensics acquisition of IMVU: a case study. *Journal of Association of Digital Forensics, Security and Law, 10*(4), 69–78 (10 pages). https://doi.org/10.15394/jdfsl.2015.1212

6. Eliot, T. S. (1934). The Rock, Hardcourt, Brace and Company Inc. https://archive.org/details/in.ernet.dli.2015.3608/mode/2up

7. Celko, J. (1999). *Data and databases: Concepts in practice, Morgan Kaufmann* (1st ed.). ISBN-13: 978-1558604322.

8. Connolly, T., & Begg, C. (2010). *Database systems: A practice; approach to design, implementation and management* (5th ed.). Addison Wesley.

9. Coronel, C., et al. (2020). *Database principles: Fundamentals of design, implementation and management* (3rd ed.). Cengage Learning EMEA. ISBN-13: 978-1473768048.

10. Assaf, W., et al. (2018). *SQL server 2017 administration inside out* (1st ed.). Microsoft Press. ISBN-13: 978-1509305216.

11. Freeman, R. G. (2013). *Oracle database 12c new features* (1st ed.). McGraw Hill.

12. Matthew, N., & Stones, R. (2005). *Beginning databases with PostgreSQL: From novice to professional* (2nd Corrected ed., Corr. 3rd printing ed.). A Press. ISBN-13: 978-1590594780.

13. Grippa, V. M., & Kuzmichev, S. (2021). *Learning MySQL: Get a handle on your data* (2nd ed.). O'Reilly Media, Inc. ISBN-13: 978-1492085928.

14. Coronel, C., & Morris, S. (2018). *Database systems: Design, implementation, & management* (13th ed.). *Cengage learning* (13th ed.). ISBN-13: 978-1337627900.

15. Rockoff, L. (2016). *The language of SQL* (2nd ed.). Addison-Wesley Professional.

16. Kreibich, J. A. (2010). *Using SQLite* (1st ed.). O'Reilly Media. ISBN-13: 978-0596521189.

17. Aouad, L., Kechadi, M.-T., Trentesaux, J., & Le-Khac, N.-A. (2012). An open framework for smartphone evidence acquisition. In P. Gilbert & S. Sujeet (Eds.), *Advances in digital forensics VIII* (pp. 159–166 (8 pages)). Springer Berlin Heidelberg. https://doi.org/10.1007/978-3-642-33962-2

18. Thantilage, R., & Le-Khac, N.-A. (2019). Framework for the retrieval of social media and instant messaging evidence from volatile memory. In *18th IEEE International Conference on Trust, Security and Privacy in Computing and Communications (IEEE TrustCom-19)* (CORE Rank A).
19. Cents, R., & Le-Khac, N.-A. (2020). Towards a new approach to identify WhatsApp messages. In *19th IEEE International Conference on Trust, Security and Privacy in Computing and Communications (IEEE TrustCom-20)* (CORE Rank A).
20. Wijnberg, D., & Le-Khac, N.-A. (2021). Identifying interception possibilities for WhatsApp communication. *Forensic Science International: Digital Investigation, 38*(Supplement), 301132. https://doi.org/10.1016/j.fsidi.2021.301132

Chapter 3
Signal Instant Messenger Forensics

Shuo Yan, **Kim-Kwang Raymond Choo**, and **Nhien-An Le-Khac**

3.1 Introduction

The rapid development of personal computing devices and social media has fundamentally influenced the living model of our world. The traditional short-text message, phone call and email are no longer enough for the current social life. The gravity of modern social life has moved from reality to a virtual dimension: desktop, laptop, tablet and mobile phone. The boundary has become vague thanks to the "never offline" networking. The explosive increase of online duration on mobile devices and the diverse instant-messaging and social media applications are strong proof of the new tendency. The instant-messaging application allows people worldwide to communicate in real-time, and share the moments of their lives with friends in text, audio, video, and multi-media formats. WhatsApp, the most popular mobile instant messenger application global, enjoys more than 2000 million monthly active users.

Being the products that cover the detailed aspects of users' social life, the security and privacy of instant-messaging applications have been a controversial topic in recent years. The call for transparent terms, stronger encryption mechanisms, and no-sharing of user data to authorities from the user community has driven the instant-messaging application manufacturers to reinforce the security and privacy settings of the products. Nowadays, most popular IM applications like WhatsApp, Signal, Viber, and Telegram have equipped with End-to-End (E2E) encryption of messages, audio and video messages, audio and video calls, media shared, and even metadata [1]. Also, applications benefit from mobile device security configurations like app-lock, screen lock, PIN and multi-factor authentication (MFA) which are widely adopted on iOS and Android devices. However, the anxiety of users towards the potential privacy risks is never eased. The concern was again triggered by WhatsApp's updated terms of conditions, the instant-messaging market leader owned by Facebook[1] (Converted

---

[1] Meta (2021). The Facebook Company Is Now Meta. [online] Meta. Available at: https://about.fb.com/news/2021/10/facebook-company-is-now-meta/.

to META in 2021), the social media tycoon at the beginning of 2021. WhatsApp users received a popped-out notification of the new privacy policy going into effect in February 2021 (extended to May 2021), which no longer includes the option to allow users to refuse data sharing with parent company Facebook. "Instead, the new policy expressly outlines how WhatsApp will share data (stuff like user's phone number, profile name, and address book info) with Facebook" [2]. Although WhatsApp issued a public explanation that the major changes and data sharing were for the business users, and WhatsApp would continue to respect the user privacy, the acceptance of the updated terms was mandatory for users to continue using all WhatsApp functions. Signal, an alternative instant messenger of WhatsApp, established in 2014 and owned by the non-profit Signal Foundation, received an explosive growth in new user registration following the WhatsApp Terms update [3, 4].

Since it was founded, Signal messenger [5], the open-source application, was a major player in the instant messaging market with a reputation for its outstanding security and privacy. The endorsements from formal NSA contractor Edward Snowden and Elon Musk made Signal the new rising star. In May 2017, the Signal was approved by the Senate Sergeant at Arms for U.S. Senate staff to use [6]. In 2020, when interviewed by Reuters, a U.N. spokesman mentioned that "The senior officials at the U.N. have been instructed not to use WhatsApp, it's not supported as a secure mechanism [7]" Meanwhile, Signal has been adopted as a UN standard for the exchange of sensitive content [8]. Also, in February 2020, European Commission staffs have been communicated internally, encouraging them to switch to Signal messaging service [9, 10].

On the other hand, the excellence of Signal's security mechanism and privacy configuration has become a new obstacle for digital forensic investigators. Like its market competitor Telegram, which has been used by organized crime groups and perpetrators of illegal activities, Signal could be abused for inappropriate uses to hide the digital traces of malicious online activities. The various options to delete messages and shared media shared within Signal and the little digital trail left in the forensic extraction create many difficulties for investigations relying on instant messaging evidence. The independence from any cloud storage or social media makes Signal a "stand-alone black box" for the digital forensic community. The End-to-End encryption of all data and metadata within Signal makes interception an even harder task, not to mention that Signal does not store data except basic user identity in their servers. The first necessary step to investigate cases involving Signal messenger is to under it. Therefore, the digital forensic community needs to understand how Signal builds its security mechanism and find options to bypass the technical obstacles between investigators and Signal data. Then, explore the options to overcome the difficulties and even use the security features to achieve investigation goals. These two steps can be combined to establish an operational framework that can guide investigators from the very beginning of the investigation to the final report. In the past years, mobile forensics has become a hot topic, especially the analysis of instant messaging applications like WhatsApp, Viber, Skype, and Kakao Talks. Nevertheless, research on Signal is much less compared to the attention on its major alternatives. The existing Signal research mainly focuses on post-mortem forensic acquisition and analysis of the digital artifacts parsed or carved by forensic software from a single platform, such as iOS, Android or Signal Windows. Also, the post-mortem forensic research on Signal fails to discuss the Signal database structure and how the messages are deleted from Signal messenger. On the other hand, there is no research to examine the cross-platform message synchronization of Signal and how to leverage the multiple linked devices to retrieve Signal data. There is also a gap in research from an OSINT and real-time data capturing perspective. This chapter proposes an integrated approach to guide digital forensic investigations of Signal messenger.

3.2 Basic Features

3.2.1 Signal Messenger

Signal messenger is a cross-platform encrypted instant messaging application developed by the non-profit Signal Technology Foundation and Signal Messenger LLC. Signal currently covers iOS and Android on mobile devices and Windows, macOS and Linux on desktop devices. Signal software is free and open-source. The clients are published under the GPLv3 license, and the server code is published under the AGPLv3 [11]. Users can register a Signal user account with a standard cellular telephone number or virtual telephone number, e.g., Google voice. Communications in Signal is end-to-end encrypted, including messages (text, audio and video), one-to-one audio/video call, group chats and calls, and shared media. Signal is famous for its encryption protocol—Signal Protocol (TextSecure Protocol).

3.2.2 Disappearing Messages

Disappearing message is the feature that allows Signal users to define a "life-duration" of messages [12]. A timer is defined with disappearing messages, and Signal messenger will remove the messages from a user's device when the timer elapses. Disappearing messages can be enabled and managed by any participant in the chat, and the timer setting applies to all new messages after the timer has been defined or modified. As Signal supports multiple platforms, it synchronises the timer modification on all devices linked to the Signal account. For the message sender, the timer starts after the message is sent, while on the receiving side, the timer starts after the recipient reads the message. No or little trace was left on the device. A similar feature is also available in WhatsApp and Viber, but the previous version of WhatsApp was not deleting the messages from its database [13]. Given that major mobile and desktop OS allow users to receive message previews as notifications, the notification could be used as an approach to read Signal messages while not changing the message read status [14]. Digital forensic investigators could benefit from the notifications; however, Signal has a privacy feature to disable the preview of message content in the notification. This part will be covered in Chap. 4.

3.2.3 Delete for Everyone

Signal allows the message sender to delete a sent message from all devices that received the message in the specific chat [15]. WhatsApp also provides a similar feature to users [16]. Unlike WhatsApp offering one-hour windows for deleting for everyone, Signal extends the windows to three hours. Also, Signal explicitly includes the words that quoted messages will not be deleted in its official documentation, while WhatsApp fails to notify its users of this widely used bypass technique [17]. Also, WhatsApp mentioned in its documentation that shared photos that have been saved to the device by the recipient may not be deleted, while Signal does not cite any exception regarding the shared media. Moreover, WhatsApp provides users with a notification whether the deletion for everyone is successful, but Signal marks the message with "you deleted this message".

3.2.4 View-Once Media

At the beginning of 2020, Signal introduced a unique feature: view-once media [18]. It allows users to configure the shared photos and videos to be removed automatically from a chat after the recipient has viewed the media. This feature is only available on Signal iOS and Android, but not the desktop client. The maximum duration of view-once media is 30 days after the media is sent, and the term will be shorter if the recipient opens the media or it has been set as a disappearing message [19]. WhatsApp is also catching up with its competitor to introduce a similar feature on the latest version of WhatsApp Android beta [20].

3.2.5 Mark as Unread

Mark as unread is a feature available in Signal mobile clients as well as desktop clients. It allows users to mark a specific chat as unread [21]. WhatsApp has a similar feature, and it only marks the message from the recipient side but does not change the message read status from the sender side. Signal does not explicitly mention how this feature works with the disappearing messages, delete for everyone and view-once media. This part will be discussed in Chap. 4.

3.2.6   Show in Suggestions

Show in Suggestions is a unique feature only available in the Signal iOS client. When an iOS device user enables the "Suggestions while Searching, Suggestions on Home Screen", "Suggestions When Sharing" options, and the Show in Suggestions feature in Signal; the iOS device will show a list of recent Signal contacts when sharing from other apps.

3.2.7   Backup and Restore Messages

Unlike WhatsApp and other instant messengers, Signal does not support cloud backup solutions. According to Signal official documentation, Signal messages, pictures, files, and other contents are only stored locally on the user device [22]. Therefore, message restoration is not supported in Signal or the account transfers. Signal supports data transfer between different Android devices and restores account information; besides, Signal Android allows users to restore data from a manually created backup file. A backup file is protected with a 30-digit passphrase, which cannot restore from the backup if forgotten. An alternative solution is to create a new backup file with a newly generated 30-digit passphrase, which is currently being used by mobile forensic software like Oxygen for Signal data extraction [23]. On iOS client, Signal only support data transfer between two iOS devices with Signal installed. The authentication is via QR code scanning. This could be considered an option for digital forensic investigators of law enforcement to take over the subject's Signal account and receive real-time information, such as group chats.

3.3   Related Work

Mobile forensics, especially research on forensic analysis of instant messaging applications, has been a hot topic. One important reason is that modern social life depends on communicating through chats on mobile devices and the cross-platform user experience. With the COVID-19 pandemic, face-to-face communication has been replaced by instant messaging and video chats. Nevertheless, there is a gap in research specifically on Signal messenger, compared to the diverse research on WhatsApp, Viber, WeChat and other instant messengers. Therefore, this chapter is developed on the research of similar instant messengers of Signal.

Alissa and al. conducted a comparative study on four digital forensic software specifically used to extract data from WhatsApp. In the research paper, the authors clearly defined the importance of WhatsApp forensics in criminal investigations and explained the database structure and the data storage of WhatsApp on Android devices [24]. Also, the authors designed a framework to test WhatsApp forensic software by combining the NIST Mobile Device Tool Test Assertions and the Test Plan and researchers-specified criteria. The author tested Guasap forensics, Elcomsoft WhatsApp Explorer, WhatsApp key/DB Extractor, and SalvationData WhatsApp Forensics tool. The comparative study indicated that the four WhatsApp forensic tools could extract WhatsApp data like messages and contacts, but not all WhatsApp forensic tools tested support extracting WhatsApp media, recovering deleted WhatsApp data, and hashing the extracted data. Also, three of the four tools extract WhatsApp data via downgrading the WhatsApp version, which interacts directly with WhatsApp user data and remains a risk to impact the data extraction. On the other hand, the research does not include important criteria for WhatsApp forensics on Android devices, for example, a comparative study of the data extracted from rooted devices against the unrooted device. In addition, the research does not include a database-level analysis of the data extracted by the four software tested. Another limitation of the study was that WhatsApp iOS, WhatsApp for Web and WhatsApp Desktop were not included. Also, the research only had four software specifically designed for WhatsApp forensics, but not the one-stop mobile forensic software like Cellebrite UFED, Oxygen Forensic Detective, which can extract WhatsApp data and other related user data from mobile devices.

Rick Cents et al. used the network packet forensic wiretap and analysis to identify the indicators of WhatsApp message communication between different Android devices [25]. The research provides law enforcement with an alternative solution when the physical device is not seized. Analyzing the network wiretap data shows that sending and receiving WhatsApp on an Android device will form a pattern of network communication between the sending/receiving device and the WhatsApp server. The patterns could be used to determine if a target was communicating using WhatsApp within a certain time. However, the research was conducted only on Android devices, without including iOS devices, WhatsApp Desktop and WhatsApp for Web. Also, the pattern found was limited to one-to-one chat via messages. The research does not include audio messages, media and files shared, and WhatsApp calls.

Marshall wrote a case note in which he conducted a brief test on the URLs shared within WhatsApp on the mobile handset [26]. Marshall tested the URLs pointing to online video files, which are then shared in WhatsApp chats. His test indicates that online videos may remain available for direct downloading even if the associated messages/chat session had been deleted from the WhatsApp database of the sender/receiver. Marshall's tests inspired an alternative approach to determine and access media files shared in WhatsApp chats other than image thumbnails or video preview frames. However, the test was in a starting phase which did not include a detailed analysis of what types of media files can be identified and the availability of direct downloading of different web servers that popular video files are hosted. Also, the research fails to discuss if the media shared in WhatsApp has any artifacts pointing to the WhatsApp servers.

The research paper of Choi et al. discussed the backup database encryption mechanism and proposed an offline password cracking methodology to decrypt the backup database of KakaoTalk [27]. The research is limited to Windows PC desktop clients but not the mobile app. Also, the research does not include how the KakaoTalk message and file shared synchronizing between the mobile app and the desktop client.

Ichsan and Riadi tested the capability of four different digital forensic software to extract data from IMO messenger from Android mobile devices [28]. The research compared the data extracted from both rooted and non-rooted Android devices, and the test results showed that MOBILedit could not create a physical image of the non-rooted Android device. On the other hand, IMO messenger data, including account information, messages, and media files, including the deleted messages, can be extracted on a rooted Android device. The research was conducted following the DFRWS method to ensure that the process is forensically sound. However, the research did not include a detailed analysis of the data extracted from the digital forensic software used; especially only one physical image of the mobile testing devices was created with MOBILedit without a comparative analysis. Furthermore, the research did not include an analysis of the database structure of how IMO messenger Android client stores data. The data analysis is only limited to the automatic parsing and carving stage, but no manual analysis is included.

Karpisek and al. experimented on WhatsApp communication between WhatsApp clients installed on Android devices and the WhatsApp server to understand the call signalling messages of WhatsApp audio calls [29]. The researchers captured the network traffic of a WhatsApp audio call using Wireshark with WhatsApp dissector. By decrypting and analysing the pcap file, the researchers closely examine the authentication process between WhatsApp clients and the WhatsApp server (full handshake when initially establishing the communication, and then half handshake with previously generated session key). The researchers also revealed that WhatsApp used Opus 8 or 16 kHz sampling rates for voice media streams. The researchers also decrypted the client and relay server IP addresses used during the communications. The research proves that by capturing and analysing the WhatsApp network traffic data, forensic investigators can retrieve artifacts like the phone number of call participants, the call establishment and termination timestamp, the call duration, the call voice codex, and the IP addresses of both involving clients and all relay servers used. However, the research limits for WhatsApp on Android devices but failed to include WhatsApp on iOS devices, WhatsApp desktop and WhatsApp Web. Therefore, it would be useful to understand if WhatsApp uses an identical or similar authentication process on other platforms. Moreover, the research could include the analysis of network communications between WhatsApp mobile clients and desktop clients to understand whether it is possible to intercept the data synchronization process.

Conti and al. conducted a comparative forensic analysis of three popular VoIP messengers, Viber, Skype, and WhatsApp, on the Android platform [30]. The research focused on analysing the data storage directories of the three target messengers within the Android system and comparing what data can be recovered or found from a logical extraction of the handset. Conti and al. indicated the file system directories Viber, Skype, and WhatsApp located and what data can be extracted from the SQLite databases. The comparative analysis showed that Viber and WhatsApp have a similar database structure, and the data can be extracted. Skype, on the other hand, leaves more forensic artifacts like a client IP address. Nevertheless, the research failed to compare the data extracted via logical extraction, file-system extraction, and physical extraction of the Android handset. Also, the research test environment is limited to rooted Android devices, and it did not analyse the detailed difference between data extracted from rooted and unrooted devices.

Al-Rawashdeh et al. conducted a post-mortem forensic analysis on the Kik instant messenger on Android platform [31]. The researchers classified eight typical scenarios of Kik usage that crime perpetrators may be used to cover their digital traces. The Kik messenger, which is popular because of its user anonymity and registration without a valid mobile number. The research focused on forensically retrieving the Kik messenger messages from the data dumps of Android device NAND memory and processing heap memory. The results showed that Kik messages could be extracted from the Android memory dumps, including the deleted ones. However, the test was conducted with an Android emulator instead of a physical Android mobile device. Also, the test did not explain why certain messages were found from both memory dumps of NAND RAM and process heap memory, while others were found in only one dump. Besides, the test only focused on the messages instead of the media, user account information and other application artifacts. The research could also be extended to the iOS platform as a comparison.

Kukuh, Muchamad et al. created a logical image of an Android device with IMO messenger installed and analysed the database structure of IMO messenger [32]. Their research reveals how data is stored in the Android internal memory and what application artifacts can be found from the data structure. Furthermore, the research can be extended by analysing the database tables to understand the data model of IMO messenger and to understand if deleted messages and media could be recovered in the forensic acquisition phase with logical, full file-system and physical extractions.

Agrawal and Tapaswi conducted a thorough forensic analysis of the Google Allo instant messenger on the Android mobile device. Compared to the related research, Agrawal and Tapaswi tried to use the Android built-in functionalities and Google ADB to acquire application data from the device internal memory instead of relying on 3rd party commercial forensic software [33]. On top of the acquisition, they conducted a detailed analysis of the Google Allo application's database structure and revealed where to locate the user information, messages, media files, and other forensic artifacts that digital investigators are targeted. They also repeated the acquisition process with MOBILedit software and compared the data extracted by the commercial software and the "native" acquisition. The comparison was designed with an algorithm developed by the authors, which also considered the margin of error. Their testing result shows that the "native" acquisition can extract as much or even more application data than the commercial software while largely reducing the risk of tempering evidence intentionally or unintentionally. However, the research fails to test the proposed acquisition method with actual scenarios and to discuss why investigators should choose to acquire data with the built-in Android functionality instead of commercial forensic software.

Riadi et al. conducted a comparative analysis of the performance of different digital forensic software [34]. The testing and analysis used the "Mobile Device Tool Test Assertions and Test Plan ver. 2" and "Mobile Device Tool Specification ver. 2" published by the NIST and adopted the Blackberry messenger, one of the most popular instant messengers on the local market for Android device as the target of analysis. The research focuses on testing the core assertions of logical acquisition and optional assertions of physical acquisition of the tested forensic software. The research can be considered a reference for mobile forensic software testing and validation. However, the research fails to include a comparative analysis of different instant-messaging applications, especially those with a larger market share.

Conti et al. researched social media and instant messaging apps and web clients based on the FxOS, the mobile OS platform developed by Mozilla [35]. The contribution of their research was to separately acquire the internal phone memory image and the volatile memory images and analyse what forensic artifacts can be retrieved from both images. The research revealed that mobile application activities, user profiles, passwords and SMS verification codes could be retrieved from the volatile memory image instead of the internal mobile image. Also, the research compared the artifacts retrieved from the mobile app and the corresponding web client. However, the research focuses only on an OS platform with a relatively small user community, not reflecting the latest trend in the instant messaging market.

Judge conducted physical and logical extraction on Android and iOS mobile devices with Signal messenger installed and analysed the application artifacts retrieved by commercial and open-source forensic software [36]. Judge's research revealed that on non-rooted/non-jailbroken mobile devices, neither commercial software nor open-source forensic software could retrieve much application data from Signal messenger. The research proves the security of Signal and raises more questions to the digital forensic community on how to acquire Signal messenger data in a forensically sound approach. However, the research fails to discuss Signal desktop or to include an in-depth analysis at the database level to understand the database tables, what traces of the deleted messages can be found from the Signal database and how to decrypt the Signal database. Instead, only the parsed results of forensic software were discussed, but no manual analysis of the database tables or a comparison of data retrieved from different types of acquisitions is included.

Wijnberg and Le-Khac systematically designed a framework for law enforcement to acquire, analyse and report WhatsApp data in the forensically sound approach [37]. More importantly, the author focused on testing the framework in various scenarios for law enforcement officers to intercept the WhatsApp communication to generate real-time or in-time intelligence. The tests conducted by the author included intercepting WhatsApp calls, taking over WhatsApp accounts, intercepting via WhatsApp Web, and gathering WhatsApp user data via OSINT research. The research provides insights for law enforcement to acquire real-time WhatsApp data, despite the difficulties with end-to-end encryption adopted by WhatsApp and other major instant-messaging tools. However, WhatsApp Desktop is not included in the research. Also, the research does not combine the in-time intelligence, OSINT research and forensic examination into an integrated framework.

As covered in Sect. 1.1, there is a gap in research on Signal instant-messaging applications. Compared with the research conducted on WhatsApp, Telegram and other major instant-messaging applications, the research specifically on Signal is relatively limited. Although commercial digital forensic software supports the extraction of Signal data, little research discusses the application data storage of Signal mobile apps and desktop clients in detail. Especially what artifacts can be founded in Signal clients and what deleted artifacts could be recovered. Although, Signal does not support cloud backup, and no data is stored in its server except basic user account information. This security mechanism reduces the possibility of wiretapping the Signal network communication or accessing data via cloud storage. Also, the digital forensic investigators need to receive real-time data from Signal, given that the features like view-once media, disappearing messages, and delete for everyone also contribute to the coverage of digital trace. Moreover, digital forensic investigators need to understand what data related to Signal messenger can be obtained without having the physical device seized. This chapter will answer these questions and lead toward a new framework as a reference for digital forensic investigators to handle Signal data.

3.4 Forensic Methods

3.4.1 Experimental Platforms

To discuss the proposed framework in this chapter, the author established a testing platform. The detailed breakdown of the platform is in Table 3.1.

The testing platform covers Signal desktop clients on both Windows and macOS and Signal mobile applications on iOS and Android. The three iOS devices are included, among which the iPad Air and iPhone 7 Plus will be linked to the same Signal account to test the data synchronization between linked devices, as well as to test the Signal data transfer between iOS devices.

Due to the limitation of hardware resources, the Android devices will be virtualized by the Android emulator BlueStacks 5. Two Android devices will be included in the testing platform. Regarding the software, Oxygen Forensic Detective and Cellebrite UFED Touch 2+ Physical Analyzer will be used to extract data from testing mobile devices in a forensically sound approach. Magnet AXIOM will be used to acquire Signal data from Windows and macOS computers and for forensic analysis.

The testing platform mainly generates test data of Signal one-to-one chat between two mobile devices and the data synchronization between linked devices. Therefore, the detailed testing devices are paired, as shown in Fig. 3.1.

Table 3.1 Testing platform

Category	Platform	Operating system	Model	Specification
Hardware	Desktop	Windows 10 Pro 64-bit version: 19043.1083	HP OMEN 15-dc0850nd laptop	Intel(R) Core (TM) i7-8750H CPU @ 2.20 GHz 2.21 GHz 32.0 GB RAM Memory
Hardware	Desktop	macOS Big Sur	MacBook Pro (Retina, 13-inch, Late 2013)	2.4 GHz dual-core Intel Core i5 processor 8 GB of 1600 MHz DDR3L onboard memory
Hardware	Mobile	iOS 14.6	iPad Air (4th Gen)	A14 Bionic chip 256 GB flash memory
Hardware	Mobile	iOS 14.6	iPhone 8	A11 Bionic chip 256 GB flash memory
Hardware	Mobile	iOS 14.3	iPhone X	A11 Bionic chip 256 GB flash memory
Hardware[a]	Mobile	Android	Samsung Galaxy S20 Ultra	4 Cores 4 GB RAM
Hardware	Mobile	Android	Samsung Galaxy S10	4 Cores 4 GB RAM
Software		Windows	Magnet AXIOM	
Software		Windows	Oxygen Forensic Detective v13.6	
Hardware		Windows	Cellebrite UFED Touch 2 7.45.1	
Software		Windows	Cellebrite Physical Analyzer 7.46	

Due to limit of resource, the Android mobile handsets are virtualized using BlueStacks Android emulator on Windows

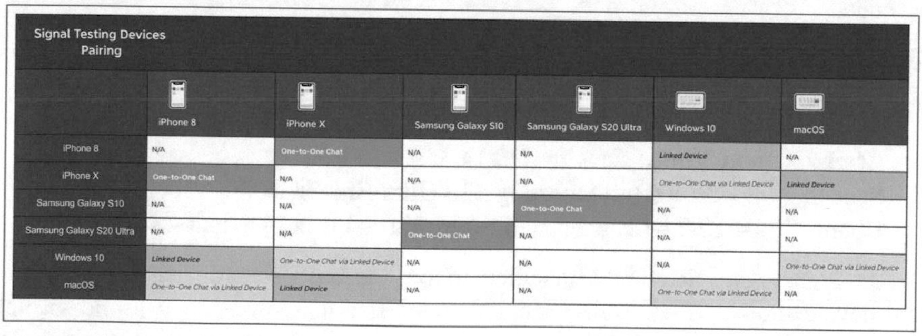

Signal Testing Devices Pairing	iPhone 8	iPhone X	Samsung Galaxy S10	Samsung Galaxy S20 Ultra	Windows 10	macOS
iPhone 8	N/A	One-to-One Chat	N/A	N/A	Linked Device	N/A
iPhone X	One-to-One Chat	N/A	N/A	N/A	One-to-One Chat via Linked Device	Linked Device
Samsung Galaxy S10	N/A	N/A	N/A	One-to-One Chat	N/A	N/A
Samsung Galaxy S20 Ultra	N/A	N/A	One-to-One Chat	N/A	N/A	N/A
Windows 10	Linked Device	One-to-One Chat via Linked Device	N/A	N/A	N/A	One-to-One Chat via Linked Device
macOS	One-to-One Chat via Linked Device	Linked Device	N/A	N/A	One-to-One Chat via Linked Device	N/A

Fig. 3.1 Signal testing device pairing

3.4.2   Datasets

The testing dataset used in this chapter is created in a combination of the following two datasets, which are both publicly available for academic usage:

- Chat logs dataset (1100 chat logs) created by Tarique Anwar and Muhammad Abulaish for their article "A social graph-based text mining framework for chat log investigation." Published in the proceedings of Digital Investigation Journal Volume 11, Issue 4, 2014.
- DFRWS 2006 Challenge dataset containing various files including JPEG, ZIP, HTML, Text, Microsoft Office file, MP3, MPG, and PDF.

The testing dataset contains one-to-one text messages, imagery content shared, URL shared, Microsoft Office documents, archived files, and local audio and video files. The testing dataset is constructed to simulate a genuine Signal one-to-one chat, and the testing dataset will cover Signal features like disappearing messages, delete for everyone, view-once media, etc. The chat-style dataset is shown in Fig. 3.2 (3 parts-crossed over on 3 pages).

3.4.3   Forensic Scenarios

3.4.3.1   Signal Account Take Over

Pre-requisites

- Signal messenger is installed on the handset of the target of the investigation.
- The target's SIM card, whose number was used for Signal registration, is cloned, or the SMS traffic is wire-tapped (near) real-time to receive the Signal verification code.

Test Environment

The test environment consists of two Android virtual devices:

1. iPhone X (bob iPhone X) with iOS 14.3;
2. Samsung S20 Ultra (Bob Samsung S20 Ultra BlueStacks);
3. Ubuntu 18.04 LTS 64 Bits virtualized by Virtual Box in Windows 10.

The Signal Android client is installed on the Android virtual device and the physical device iPhone X. A Signal account is registered first on the Android virtual device, and then the author tries to register a new Signal account using the mobile number linked to the virtual device on iPhone X. The Signal Linux client is installed in the Ubuntu virtual machine and linked to the Signal account registered with the virtualized Samsung S10.

Part I

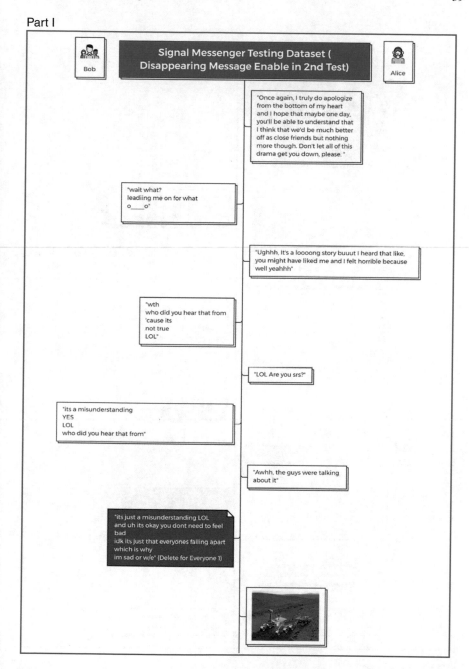

Fig. 3.2 The testing dataset—Part I (in chat), Part II (in chat), and Part III (in chat)

Part II

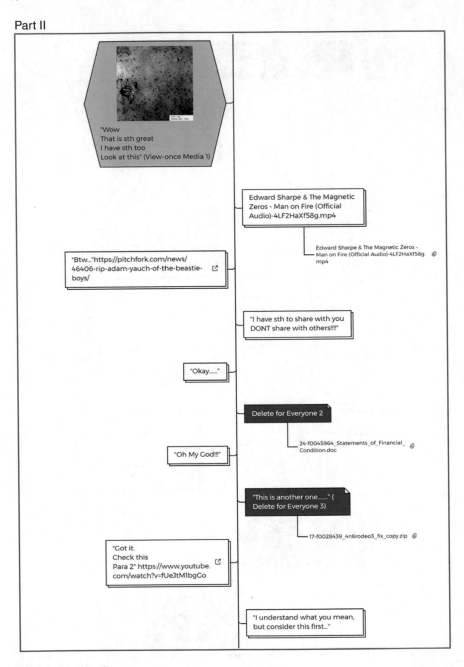

Fig. 3.2 (continued)

Part III

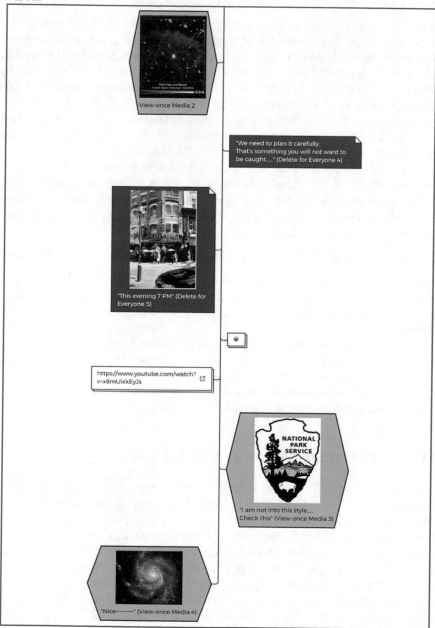

Fig. 3.2 (continued)

Test Scenario 1—Signal Registration Lock Is Enabled on the Target's Handset

In this test scenario, the Signal security feature registration lock is enabled. This means registering Signal with the same phone number on a new device requires entering the user-defined PIN code, and this step could not be skipped. This security mechanism stops hackers from taking over a legitimate Signal user's account by wiretapping the victim's SMS traffic. However, this mechanism also increases digital forensic investigators' difficulties in taking over the target's Signal account from a technical perspective. This test aims to understand how the Signal registration lock will impact digital forensic investigators from taking over a target's Signal account by registering a new Signal account with the target's phone number.

A new Signal account will be registered on the iPhone X running iOS 14.3. Then, the registration lock is enabled from the Signal settings. Then, the author tries to register a new Signal account with the same phone number used by Signal on iPhone X. The registration is conducted on a virtual Samsung Galaxy S20 Ultra running Android Nougat 64-bit created in BlueStacks Android emulator.

Test Scenario 2—Signal Registration Lock Is Disabled on the Target's Handset

In this circumstance, the Signal registration lock is disabled on the target's handset, which is not the default configuration of Signal. Digital forensic investigators could register Signal on an examiner's mobile handset as if they own the target's phone number.

A new Signal account will be registered on the iPhone X running iOS 14.3. Then, the registration lock is disabled from the Signal settings. Then, the author tries to register a new Signal account with the same phone number used by Signal on iPhone X. The registration is conducted on a virtual Samsung Galaxy S20 Ultra running Android Nougat 64-bit created in the BlueStacks Android emulator.

3.4.3.2 Signal Activity Monitoring with Linked Device

Signal uses the mobile handset on which the Signal account is registered as the master device. Similar to its major competitor WhatsApp, Signal allows a user to connect other devices as linked devices and create a seamless cross-platform user experience. However, as tested in the previous section, Signal account takeover, Signal does not allow a user to link a second mobile phone. However, a mobile device such as an iPad can be linked to an iPhone. Signal currently supports both iOS and Android platforms for the mobile side, Windows, macOS and Linux (Debian-based) for the desktop side. Therefore, an iOS/Android mobile phone can link to all other available platforms. Once linked, Signal chats will be synchronized on all linked platforms. The device linkage is created by scanning the QR code generated on Signal clients of different platforms with the mobile phone used for Signal registration.

In this test, one Signal account is created on the iPhone 8 and linked with the Windows 10 examiner laptop. Similarly, another Signal account is created on the iPhone X and linked with the macOS Big Sur examiner laptop. The main aim of this test is to understand if investigators cannot seize the mobile device of an investigation target, what Signal data can be obtained by seizing a linked device or linking an investigative device to the target's Signal account.

3.4.3.3 Signal Group Chat

Like other instant messaging applications, Signal also provides users with the feature to create groupsand chat, call, and share media and files between group members. However, unlike Telegram or Facebook, which have public groups that all application users can join groups without seeking approval from a group admin, Signal only provides secure groups. This means that Signal groups are private, which is a mechanism that all perspective group members must request and be approved before they becomeofficial group members. Previously criminal investigations and white-collar crimes have witnessed abuses of group features of instant messaging applications, such as illegal intelligence exchange, drug, weapon trade, etc. On the other hand, law enforcement can go undercover in the group to monitor and collect group chats as normal group members. This test aims to understand if investigators can recycle this investigation technique when facing Signal messenger.

In this scenario, the test aims to assess the possibility of joining the Signal group and monitoring the Signal group chat. The testing configuration is:

1. Virtual Samsung S10—Claire Samsung S10 SignalTest to create a new Signal group named SignalTest.

    (1) Signal provides two options when a user requests to join a Signal group: explicit approval of a request by the group admin or no explicit approval process.
    (2) Signal provides four different approaches to share a group link: Share via Signal, QR code, copy link and share with other apps.

2. bob iPhone X (linked with Signal Mac): in the first round, Bob iPhone X will join the SignalTest group with no explicit approval process via scanning QR code.
3. Virtual Samsung S20 Ultra—Joe Samsung S20 Ultra. Joe will join the SignalTest group without explicit approval via copy and manual share. The group link is copied from Claire and pasted in a Note to Self in Joe; click send, and click the link in the sent note to self to join the group.

3.4.3.4 Use Signal as Source of OSINT

Instant messaging applications are breaking the barrier between communication tools and social media. Many instant messaging applications embed social networking features, such as the tight linkage between WhatsApp and Facebook. This trend has made instant messaging applications the new gold mine of open-source intelligence research. Compared with other major instant messaging applications, Signal seems to fall behind in the social network features and provides users with fewer related features. However, Signal remains a useful source of OSINT.

Search Signal Account Via Mobile Number

This function can serve as the starting point of OSINT research on Signal. Since 2014, Signal has been working on the Private contact discovery feature, which allows a Signal user to query based on Signal username or the phone number linked to a Signal account within the application to determine whether there is a valid Signal account connected to the username or the phone number. In this test, the author will test how to search for a phone number within Signal and how it can be used for OSINT investigation.

Notify When Contact Joins Signal

Manual phone number search in Signal could be a tedious and time-consuming task. To improve search efficiency, digital forensic investigators can benefit from another Signal feature—Notify When Contact Joins Signal.

This feature is enabled by default when installing and registering for a new Signal account. Signal asks new users permission to access the mobile handset's device contact book. If the user grants Signal permission, Signal will map all contacts within the device contact book and notify the user which contacts are using Signal. Therefore, instead of querying each number under investigation, digital forensic investigators can add all mobile numbers of interest into the device's contact book and allow Signal to access it. In this test, the author will walk through the entire process of searching Signal accounts via the contact notification feature.

Signal Account Profile

Signal allows users to create a personalized account profile, including profile image, username, and account description. In this test, the author will test what Signal account information can be obtained and used for OSINT investigation.

Read Receipts and Typing Indicators

To provide a better user experience in chatting, Signal and other major instant messaging applications have an icon to show whether a sent message has been read. Also, typing indicator is enabled in Signal by default to notify users that another chat participant is typing. Unlike Skype, Microsoft Teams and other instant messaging applications, Signal does not have an indicator of online status. Instead, digital forensic investigators can use the read receipts and typing indicators together to map an activity timeline of the target. Investigators can deduct the online status based on whether the received messages have been read. The two Signal indicators can be useful if investigators can monitor Signal group chat in real-time. In this test, the author will discuss how these two indicators could be used for OSINT investigation.

Shared Media and URL Preview in Chat

Signal allows users to share URLs within chat sessions and generates a URL preview with a thumbnail and brief description of the webpage that the URL points to. The URLs shared in Signal one-to-one and group chats can provide digital forensic investigators with much information about the topics covered in the chat, and provide new leads for online investigation.

3.4.4 Forensic Acquisition and Analysis of Signal

This section of the chapter will discuss the experiment setup, the detailed testing scenarios and how to collect the testing data. Before each test, the physical and virtual mobile devices are wiped completely for a fresh installation and registration of Signal. The Signal client and its data on the desktop devices are deleted for a fresh installation and new linkage with the mobile devices.

3.4.4.1 Signal iOS

The first part of the test on Signal is conducted based on the Apple iOS platform. This section will cover the test platform composition, test parameters, and tested scenarios in detail.

Test Environment

The test on the iOS platform includes two iOS mobile handsets and two linked computer devices. The detailed test platform is listed in the following Fig. 3.3. The iPhone 8 and iPhone X used for testing are reset to factory settings via the "Erase all Content and Settings". A separate testing Apple ID is created and logged in on each iOS device to install Signal messenger from the App Store. For the Windows 10 and macOS Big Sur examiner laptops, a fresh installation of Signal clients is conducted.

Signal Test on iOS Platform					
	OS Version	Alias	Test iCloud ID	Signal Version	Linked Device
iPhone 8	iOS 14.6	Alice iPhone 8	alice_signaltest@protonmail.com	5.16.1	Alice Windows 10
iPhone X	iOS 14.3	Bob iPhone X	bob_signaltest@protonmail.com	5.16.1	Bob MacBook Pro
Windows 10 Laptop	Windows 10 Pro 64-bit	Alice Windows 10	N/A	5.10	Alice iPhone 8
MacBook Pro	macOS Big Sur	Bob MacBook Pro	bob_signaltest@protonmail.com	5.7	Bob MacBook Pro

Fig. 3.3 Signal test platform on iOS

Test Scenario 1

Alice iPhone 8 and Bob iPhone X were configured in the Signal default setting, listed as follows. Signal Windows on Windows 10 examine machine is configured with default settings, linked to the Alice iPhone 8. Signal Mac is configured with default settings linked to the Bob iPhone X. Once the test environment is set; a Signal one-to-one chat is initiated on Alice iPhone 8. The messages and media included in the testing dataset mentioned in the early sections are sent and received between the two iOS mobile handsets. The two linked examiner laptops will synchronize all messages sent and received on the corresponding iOS mobile handset in real-time.

- **Chat Settings**
  - Generate Link Previews enabled
  - Show Chats in Suggestions enabled
  - Use System Contact Photos disabled

- **Notification Content**
  - Name, Content, and Actions
  - Notify when Contact Joins Signal

- **Privacy**
  - Messaging
    - Read Receipts enabled
    - Typing Indicators enabled
  - Disappearing Messages disabled
  - App Security
    - Hide Screen in App Switcher disabled
    - Screen Lock disabled
  - Calling
    - Show Calls in Recents enabled
  - Advanced
    - Always Relay Calls disabled
  - Sealed Sender disabled.

Test Scenario 2

Alice iPhone 8, Bob iPhone X, Alice Windows 10, and Bob MacBook Pro were configured with differentiating Signal settings. The same linked device configuration is used in this test. Also, the disappearing message timer is enabled and set to five minutes. Finally, the same test dataset is used for the Signal one-to-one chat between Alice iPhone 8 and Bob iPhone X. The detailed device settings are listed as follows:

Alice iPhone 8

- Chat Settings

  - Generate Link Previews disabled
  - Show Chats in Suggestions disabled
  - Use System Contact Photos disabled

- Notification Content

  - No Name or Content
  - Notify when Contact Joins Signal disabled

- Privacy

  - Messaging

    - Read Receipts disabled
    - Typing Indicators disabled

  - Disappearing Messages

    - Default Timer for New Chats—5 min

  - App Security

    - Hide Screen in App Switcher enabled
    - Screen Lock enabled (passcode/Touch ID/Face ID each time to enter Signal)

  - Calling

    - Show Calls in Recents disabled

  - Advanced

    - Always Relay Calls disabled

  - Sealed Sender disabled.

Bob iPhone X

- Chat Settings

  - Generate Link Previews enabled
  - Show Chats in Suggestions enabled
  - Use System Contact Photos disabled

- Notification Content

  - Name, Content, and Actions
  - Notify when Contact Joins Signal enabled

- Privacy

  - Messaging

    - Read Receipts enabled

- Typing Indicators enabled
  - Disappearing Messages
    - Default Timer for New Chats—disabled
  - App Security
    - Hide Screen in App Switcher disabled
    - Screen Lock disabled
  - Calling
    - Show Calls in Recents enabled
  - Advanced
    - Always Relay Calls disabled
  - Sealed Sender
    - Show Status Icon enabled
    - Allow from Anyone enabled.

Alice Windows 10

- Notification Content
  - No name or content
  - Draw attention to this window when a notification arrives
  - Play audio notification
- General
  - Enable spell check
  - Calling
    - Always relay calls enabled
    - Play calling sounds enabled
    - Show notifications for calls enabled
    - Enable incoming calls
  - Permissions
    - Allow access to the microphone
    - Allow access to the camera
  - Disappearing messages
    - Default timer for new chats—5 min (synchronized with Alice iPhone 8).

Bob macOS Big Sur

- Theme
  - Dark
- Notification
  - Name, content, and actions
  - Play audio notification
- General
  - Enable spell check
- Calling
  - Play calling sounds
  - Show notifications for calls
  - Enable incoming calls
- Permissions
  - Allow access to the microphone
  - Allow access to the camera
- Disappearing messages
  - Default timer for new chats—off.

iOS Digital Forensic Acquisition of Testing Data

Once the test has been finalized, the author conducts digital forensic acquisitions to extract data from the above-mentioned test devices. The detailed digital forensic acquisition process is shown in the following Fig. 3.4. Both full file-system extraction and logical iTunes backup will be created for the iOS devices to compare and understand what artifacts can be found from the extractions. Alice iPhone 8 and Bob iPhone X were connected to the examiner's laptop with an Apple data cable, and tap "Trust this computer" on the iPhones. iTunes logical backups and full filesystem based on Checkra1n exploit extractions were conducted with Oxygen Forensic Detective 13.6.

For Alice Windows 10 and Bob MacBook Pro used in the tests, a live RAM extraction is conducted on both devices to understand what Signal artifacts can be found from the memory. Then, to simulate port-mortem forensic investigations, both devices were powered off following a full disk acquisition with CAINE 12 Bootable USB beta release. The main aim is to simulate the different scenarios that digital forensic investigators may encounter in a real daily investigation. In some investigations, investigators may not have an opportunity to acquire the RAM but are only given a power-off computer. In these circumstances, what can be found and recovered from a post-mortem full disk acquisition will be important for the investigation.

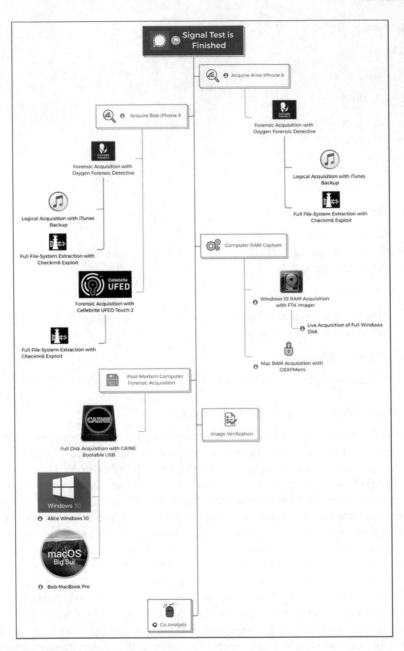

Fig. 3.4 Signal testing devices forensic acquisition process—iOS

3.4.4.2 Signal Android

This section of the chapter will discuss the detailed experiment setup, the testing scenarios, and the forensic acquisition process of Signal messenger on the Android platform.

Test Environment

The test on the Android platform includes two virtual mobile devices running Android 10 and one linked virtual desktop environment. The detailed test platform is listed in the following Fig. 3.5. Android mobile devices used in this test were created in the Android emulator BlueStacks 5. They are emulated with the Android Nougat 64-bit operating system, and the Signal Android app is downloaded from the Google Play store pre-installed in the default system image. Also, an Ubuntu Linux distribution virtual environment is created in VMWare Workstation Pro 16. Signal Linux 5.10 is downloaded and installed in the Ubuntu environment and linked to the virtual Android device Alice Samsung S10.

Test Scenario 1

The Signal test on the Android platform has a different device configuration compared with the tests conducted on the iOS platform. Alice Samsung S10 and Bob Samsung Galaxy S20 Ultra have asymmetrical Signal settings to simulate the different user experiences in real-time. The design aims to configure Alice Samsung S10 with the Signal app's strict security and privacy settings. Therefore, the features that could potentially feed the other chat participant or third-party monitors, such as read receipt, typing indicator, and URL preview are disabled. On the other hand, Bob Samsung Galaxy S20 Ultra has a "casual" style setting from a privacy perspective. Also, the disappearing message timer is enabled and set to five minutes. The detailed device configurations are listed as follows. Once the test environment is set, a Signal one-to-one chat is initiated on Alice Samsung S10. The messages and media included in the testing dataset mentioned in the early sections are sent and received between the two Android virtual devices. The linked Signal Ubuntu will synchronize all messages sent and received on the corresponding Android mobile handset in real-time.

Signal Test on Android Platform					
	OS Version	Alias	Test Google ID	Signal Version	Linked Device
Samsung S10 (Virtualized by BlueStacks)	Android Nougat 64-bit	Alice Samsung S10	kariml23jordan@gmail.com	5.18.5	Alice Ubuntu
Bob Galaxy S20 Ultra (Virtualized by BlueStacks)	Android Nougat 64-bit	Bob Samsung Galaxy s20 Ultra	kariml23jordan@gmail.com	5.18.5	N/A
Ubuntu (Virtualized by VMWare Workstation Pro)	Ubuntu 20.04.2 LTS 64 bit	Alice Ubuntu	N/A	5.10	Alice iPhone 8

Fig. 3.5 Signal test platform on android

Alice Samsung Galaxy S10

- Chat Settings

  - Generate Link Previews disabled
  - Show Chats in Suggestions disabled
  - Use System Contact Photos disabled
  - Use system emoji disabled
  - Chat backup disabled

- Notification enabled

  - Repeat alerts—Never
  - Show no name or message
  - Priority—High
  - Calls Notification enabled
  - Notify when Contact joins Signal disabled

- Privacy

  - Messaging

    - Read Receipts disabled
    - Typing Indicators disabled

  - Disappearing Messages disabled
  - App Security

    - Screen security enabled
    - Screen Lock disabled in VM
    - Incognito keyboard enabled

  - Advanced

    - Always Relay Calls enabled
    - Signal messages and calls enabled

  - Sealed Sender disabled.

Bob Samsung Galaxy S20 Ultra

- Chat Settings

  - Generate Link Previews enabled
  - Use system emoji enabled
  - Use System Contact Photos disabled
  - Chat backup enabled
  - Use System Contact Photos disabled
  - Chat backup enabled

- Notification enabled

  - Repeat alerts—Never

- Show Name and message
- Priority—High
- Calls Notification enabled
- Notify when Contact joins Signal enabled

- Privacy

  - Messaging

    - Read Receipts enabled
    - Typing Indicators enabled

  - Disappearing Messages

    - Default Timer for New Chats—disabled

  - App Security

    - Hide Screen in App Switcher disabled
    - Screen Lock disabled
    - Incognito keyboards disable

  - Advanced

    - Always Relay Calls disabled
    - Signal messages and calls enabled

  - Sealed Sender

    - Show Status Icon enabled
    - Allow from Anyone enabled.

Alice Ubuntu LTS 20.04.2

- Theme—Dark

- Notification

  - Name, content, and actions
  - Draw attention to this window when a notification arrives
  - Play audio notification

- General

  - Enable spell check

- Calling
- Always relay calls

  - Play calling sounds
  - Show notifications for calls
  - Enable incoming calls

- Permissions
  - Allow access to the microphone
  - Allow access to the camera

- Disappearing messages
  - Default timer for new chats—5 min (synchronized).

Android Digital Forensic Acquisition of Testing Data

Like the iOS tests, formal digital forensic acquisitions were conducted to extract data from testing devices. However, compared with the forensic acquisition of physical devices, the process of virtual machine acquisition is different. The Android BlueStacks virtual disk images in VDI format can be found in system path C:\ProgramData\BlueStacks_nxt\Engine. These images are collected for further forensic analysis. A snapshot file in VMEM format is created with the Signal Linux app opened regarding the Ubuntu virtual environment. The virtual memory image and the virtual disk image in VMDK format are collected for further forensic analysis.

3.4.5 Forensic Analysis

Oxygen Forensic Detective is used to create both iTunes logical backups and full filesystem extractions of the iOS devices. Besides, a full filesystem extraction of Bob iPhone X is conducted using Cellebrite UFED Touch 2. Given that the Signal Foundation claimed that Cellebrite UFED has security vulnerabilities that may impact the integrity of the digital forensic acquisition process, this chapter uses Oxygen Forensic Detective as the main mobile forensic software. However, Cellebrite products are also used for cross-validation purposes. Once the mobile forensic images are created, they are imported into Oxygen Forensic Detective for forensic analysis. The full filesystem extraction created with Cellebrite UFED Touch 2 will be imported to Cellebrite Physical Analyzer for comparative analysis.

The forensic images and memory capture from Alice Windows 10 and Bob MacBook Pro are processed by Magnet AXIOM. The virtual memory and disk images collected from BlueStacks, and VMWare Workstation Pro are also processed by Magnet AXIOM.

The analysis phase consists of three parts: automatic analysis, database review and manual analysis.

(1) Automatic Analysis: rely on commercial digital forensic software to parse data from the forensic images and carve the deleted data. Review the analytical results delivered by the forensic software. This is usually the first step of digital forensic analysis in real investigations. The automatic analysis results can provide investigators with direct information and data aggregation to save time and effort for manual analysis.

(2) Database Review: in the second step of forensic analysis, the Signal database structure review is conducted to understand where to locate Signal application data within a device's filesystem. Besides, reviewing the Signal database structure will provide investigators with information on what artifacts can be found from the Signal database and how to handle the deleted data.

(3) Manual Analysis: As the final stage of forensic analysis, manual analysis is the process for investigators to deeply analyse specific digital artefacts identified in the two previous steps.

3.5   Findings and Discussion

3.5.1   Signal Account Take Over

3.5.1.1   Test Scenario 1—Signal Registration Lock Is Enabled on the Target's Handset

On the Bob iPhone X, where a new Signal account is created, the registration lock feature is enabled under the Signal account setting. In this test, the author assumes that the user-defined PIN is obtained by digital forensic investigators, which is a pre-requirement to continue the test.

Then, on the virtual Android device Bob Samsung Galaxy S20 Ultra, a new Signal account registration is initiated with the same number used by the Signal account on Bob iPhone X. When requested the PIN code, the obtained PIN code of Bob iPhone X is input. A new Signal account is successfully created on Bob Samsung Galaxy S20 Ultra. However, the author receives a notification shortly on Bob iPhone X that his/her handset is no longer registered because the phone number was used to register Signal on a different device, as shown in Fig. 3.6. The de-registered device will no longer be able to send or receive any new Signal message. The notification will alert the target that something wrong happens to his/her Signal account. The target of the investigation may take the account back by re-registering the same Signal account on the original device (iPhone X in our test) with the same PIN code. Then, create a new PIN code to stop investigators from further attempts. In a worse situation, the target may abandon the Signal account under investigators' radar by deleting the Signal account. Therefore, taking over the Signal account with a registration lock can be risky with a low success rate.

According to Signal documentation, a PIN code will expire after seven-day inactivity. Theoretically, investigators can seize the target's mobile device, wait for seven days, and then take over the Signal account. However, if not under control, the target can easily de-register Signal from the seized device.

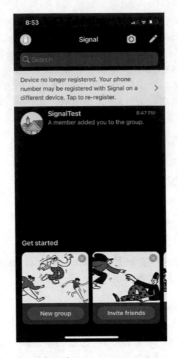

Fig. 3.6 Notification of device de-registration

Fig. 3.7 Signal Activity Synchronized in Windows 10

3.5.1.2 Test Scenario 2—Signal Registration Lock Is Disabled on the Target's Handset

When the registration lock is disabled, Signal still asks for the PIN code when registering a new Signal account on Bob Samsung Galaxy S20 Ultra, with the same number that Signal is using on Bob iPhone X. However, Signal now allows skip the PIN verification. Nevertheless, the skipping does not mean investigators could register Signal on their device without alerting the target. Instead of inputting the PIN obtained, Signal asks users to create a new PIN to overwrite the previous one when the skip option is chosen. When a new PIN is input and confirmed, the author still receives the same notification on Bob iPhone X that the device has been de-registered in Signal. Besides, the target will still receive a notification that his/her device is de-registered because the phone number is used to register Signal on another device. Again, the author can tap the notification to re-register Signal and create a new PIN to take the Signal account back.

This test shows that the registration lock mechanism of Signal remains an obstacle for the investigators to take over a Signal account without notifying the target. Furthermore, once the target is warned that he/she might be under investigation, the target may start deleting Signal data or abandon the phone number. Therefore, the Signal account takeover should be conducted with extra caution and only considered one of the last solutions.

3.5.2 Signal Activity Monitoring with Linked Device

In this test, Alice iPhone 8 has a linked device Alice Windows 10, and Bob iPhone X has a linked device Bob MacBook Pro. A computer with a corresponding Signal client installed can be linked to one and only one mobile handset with Signal installed. When launching Signal Windows or Mac, the client generates a QR code. Using the mobile camera to scan the QR code can link the computer and the mobile device.

The Signal user can review all linked devices in the "Linked Devices" list. The existing linked device can be removed from the list, and the Signal client on the linked device will receive a notification reminding the user of the linkage removal. The link relation will need to be renewed by scanning a new QR code after a period, which is the main obstacle for investigators in monitoring the target's Signal activities and receive real-time Signal messages. However, the linked devices remain one of the useful options for forensically examining the target's digital activity.

To monitor the target's Signal activity via linked devices, investigators can either seize an existing linked device or periodically scan the QR code. Regarding the artifacts that can be found and recovered from a post-mortem forensic examination of the linked Signal device, the details will be covered in a later section. This section focuses on the real-time monitoring of Signal activity.

Based on the author's test, Signal messages in text, audio, and video will be synchronized to Signal Windows, macOS or Linux client, as shown in Fig. 3.7.

Fig. 3.8 Notification of
contacts using signal

However, Signal audio and video calls are not synchronized to linked devices and
vice versa. However, this may differ for different makes and models of devices.

Interestingly, the Signal clients on the Desktop side are very similar to WhatsApp
Web, which does not download and save shared media to local disks automatically.
Instead, a download option is provided with the arrow icon beside each shared media.
Therefore, investigators need to manually save the media required when monitoring
Signal activity via a linked device. The saving action is not notified on the target's
mobile handset. On the other hand, the design of Signal desktop clients leads to
another hypothesis: the Signal desktop client will not store messages and shared
media in a database unless manually saved locally. This hypothesis will be discussed
in the forensic acquisition section.

In general, monitoring real-time Signal communication via a linked device could
be an option in the actual investigation. However, the target may be aware of the
existence of a seized linked device and remove the linkage. On the other hand, how
to scan the QR code generated by Signal desktop clients by the target's mobile phone
periodically should be the most difficult step.

3.5.3 Signal Group Chat

On the virtual Android device Claire Samsung S10, a new Signal account is created,
and then a Signal group is created with the name SignalTest. When creating a Signal
group, it is possible to include existing Signal contacts into the group; however, the
test skips this step and creates a group with only the group creator. It is worth noting
that the Signal user who creates the group is assigned the group admin role by default.

Then, Signal asks the group creator to enable and share a link to the newly created group. This is the mechanism within Signal to invite friends to join an existing group. There is an option "Approve New Members" when enabling group link sharing. This option, once enabled, will require the group admin to approve each request to join the Signal group. Otherwise, Signal users possessing the group link can join the Signal group without approval from a group admin. Both options are tested in this chapter.

In the first round of tests, the option "Approve New Members" is disabled. Then, Signal offers four different approaches to share a group link: share the group link (URL) via Signal message, generate a group link in QR code, and invite friends to scan the QR code, share the group link via third-party apps, such as email, and copy the group link for further manual sharing.

The author scans the group QR code generated on Claire Samsung S10 with Bob iPhone X, and then a notification will be shown to ask for user confirmation to join the Signal group. Once confirmed, Bob joins the group SignalTest.

Another virtual Android device Joe Samsung Galaxy S20 Ultra received the SignalTest group link via manual copy and paste and joined the SignalTest group.

Although no explicit approval process is required, all the group members will receive a notification that a new Signal user has joined the group via the group link. This notification is visible to all Signal group members. When investigators try to join the target Signal group to receive a real-time Signal group chat, the notification may alert the target of the investigation that a suspicious member has joined. However, if the Signal group has many members with high group dynamics, the notification may be flooded with new messages. Therefore, monitoring Signal group chat is still useful in real investigations.

Also, all group members of a Signal group can see the member list from Signal group settings and view other group members' profile details including profile images, descriptions, and phone numbers. This could be another valuable source of evidence for further OSINT investigation. The group admin can also remove a group member from the group, make another group member as admin, add a group member to another group, and add a group member as a contact.

Claire removes Bob and Joe from the group as a group admin at the end of the first round. The removal of the group member will also leave a notification in the group chat interface.

In the second round, Bob iPhone X and Joe Samsung Galaxy S20 Ultra will re-join the SignalTest group with the option "Approve New Members" enabled. When an explicit approval request is enabled, the group admin could grant all group members to pass the group join request or limit the authorization to only the admin. Under this configuration, Bob will join the group via the share with copy link, and Joe will join the group via communicating with Claire within Signal, and then Bob will pass Joe's request.

Firstly, enable the "Approve New Members" under the sharable link settings on Claire Samsung S10. Then, copy and paste the group link to Bob iPhone X and click the group link to request joining the group SignalTest. Then, Claire Samsung S10 as the group admin, will receive a Signal notification that there is one new request to join the group pending approval. Only the group admin can view the notification.

Once the group admin taps the notification, Signal will redirect the admin to the interface to approve or deny the pending requests. Here, Claire passes the request from Bob, and Bob joins the SignalTest group.

After Bob joined the group, modify the group permission to extend the permission to add new group members from only admin to all group members. Once applied, new requests to join the group will lead to a notification reminder of pending requests visible to the all-group members. On the other hand, the invite friends option remains only available for the group admin. However, group members can go inside the group setting to copy the group link and share it further. However, in the group setting, only the group admin, Claire, has the option to approve Joe's request. Once Bob is made an admin of the group, he can pass Joe's request. Therefore, the approval process overwrites all existing group members' permission to add new members.

3.5.4 Use Signal as Source of OSINT

Instant messaging applications are simulations of human interaction in real life; therefore, much personal information can be collected from Signal and similar social apps. This section mainly discusses how Signal features can contribute to OSINT investigations.

3.5.4.1 Search Signal Account Via Mobile Number

The feature of searching contacts by phone number can be used for OSINT investigation. If investigators have several phone numbers under investigation, the numbers can be searched within Signal to see if a Signal account is linked to the phone number under investigation. It is worth noting that Signal does not necessarily have to be registered with a valid mobile phone number. Virtual call service as Google Voice and Twilio can be used to register a Signal account. Moreover, a landline phone number is also valid for Signal account registration.

The number search function in Signal allows users to type in a phone number to be searched. If the queried phone number has no Signal account linked, the application will return an error message "Contact is not a Signal User". Otherwise, Signal will redirect the user to a new chat with the Signal account linked to the searched number. Then, the number associated with the found Signal account could be used for phone number reverse searching and social media account searching. Signal does not provide the features of searching partial phone numbers and returns a list of all matching results. Instead, only when a valid phone number is entered, and the user presses the search button can Signal show the search results. Therefore, the manual searches by phone number within Signal may be repetitive work for digital forensic investigators. Whether the search process could be automated via Signal API is beyond this chapter's scope but is considered a potential future work.

3.5.4.2   Notify When Contact Joins Signal

Notify When Contact Joins Signal can be used as an improved version of Signal account searching. This feature is enabled by default when a new Signal account is registered. Investigators can create a new contact for each phone number under investigation in the device's contact book and then launch Signal messenger on the phone. The Signal will ask for permission to access the device contact book. When granted permission, Signal will generate a blue notification reminding users that their contacts are using Signal, as shown in Fig. 3.8.

When clicking the blue notification, users will be redirected to the page to search Signal account by Phone Number. Instead of querying manually, Signal displays a list of contacts using Signal below the search pane. With this approach, digital forensic investigators can quickly focus their investigation on the Signal accounts and corresponding phone numbers in the list. This information can also be used for further OSINT investigation.

3.5.4.3   Signal Account Profile

Signal allows users to create a personalized account profile, including profile image, username, and account description.

Although Signal provides a series of pre-defined avatars and text-style avatars for the profile image, users can also use pictures from the device photo gallery and take a profile photo with the device camera. Imagery content, including user profile images, is one of the most important sources of OSINT investigation. Digital forensic investigators can adopt the profile image downloaded from publicly available online sources and conduct a reverse image search to track the target's online activity. If a Signal user chooses to use a photo containing genuine facial information as the Signal profile image, the photo can be very useful in social media investigation (Fig. 3.9).

3.5.4.4   Read Receipts and Typing Indicators

Read receipts and typing indicators together can be used to indicate Signal online status. When these two features are enabled, investigators can build a heatmap of the target's Signal online timetable, analysing the activity pattern such as when period within the day the target is actively on Signal. Especially these two indicators are useful within a Signal group chat that distinct activity patterns of each group member can be built.

However, the two indicators are not always reliable. On the one hand, both options can be disabled in the Signal setting, which will stop other chat participants to view the message read status and typing indicator; on the other hand, Signal allows users to mark a chat session as unread to remind the user him/herself for any messages of interests.

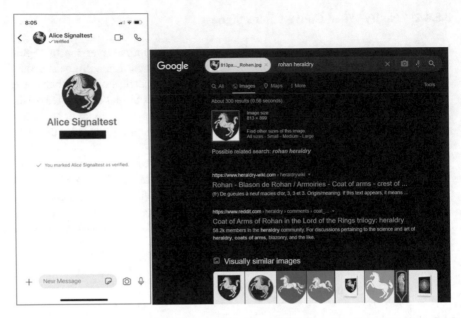

Fig. 3.9 Signal account with downloaded profile image (left); Google reverse image search of signal profile image (right)

3.5.4.5 Shared Media and URL Preview in Chat

Signal allows users to share URLs within chat sessions and generates a URL preview with a thumbnail and a brief description of the web page that the URL points to. The URLs shared in Signal one-to-one, and group chats can provide digital forensic investigators with much information about the topics covered in the chat and provide new leads for online investigation. Based on the tests conducted by the author, the URL can be previewed within a Signal chat if the message sender enables the feature. The message recipient does not need to have URL preview enabled.

3.5.5 Forensic Acquisition and Analysis of Signal

This section will focus on the test results obtained from the forensic acquisition of devices with Signal running on and explain the findings and how they can contribute to the Signal investigation.

3.5.5.1   Signal iOS

Visual Examination

When investigators have the iOS devices seized, a forensic acquisition may not be possible on site. In this case, a visual examination may be helpful for a quick assessment. In this chapter, the author focuses on four Signal features: Notification content, Hide Screen in App Switcher, and Screen Lock.

The notification content means what details can be viewed from iOS notifications. Signal allows users to choose from Name, Content, Actions, Name Only, and No Name or Content. As shown in the figure below, if Signal is configured to show name, content, and actions on iOS system notifications, investigators can visually view many incoming Signal messages and the sender information.

Screen lock is a user security mechanism; once enabled will need to enter the iOS system passcode or authenticate via biometrics to land in the Signal app. If the feature is enabled, investigators cannot view the Signal messages directly by launching the app without the passcode.

Hide Screen in App Switcher is another user privacy setting. When switching apps on an iOS device, the content on the current app view will be shown to the user. Therefore, investigators may adopt this system feature to view Signal messages. However, if the target enables the Hide Screen in App Switcher, the Signal app will only show a blue background with the Signal logo when switching apps.

iTunes Logical Backup

iTunes logical backups of Alice iPhone 8 and Bob iPhone X are created with Oxygen Forensic Detective (Fig. 3.10). The logical backups use the standard iTunes backup mechanism to backup iPhone user data including contact book, iMessage, calendar, and data from built-in apps on iPhone. As a logical backup, no deleted data and data from unallocated space will be included. The logical acquisition aims to understand what Signal artifacts can be found in an iTunes backup, which is an Apple-provided backup solution for all regular iOS device users. It is also useful for investigators to assess whether the logical acquisition should be considered an option when no other acquisition technologies are equipped.

From the following figure of the Oxygen Forensic Detective device extraction dashboard, we notice that Signal is not listed in the apps automatically extracted and parsed by Oxygen. However, this does not mean that no Signal artifacts are extracted. The next step can be a global search within the iTunes backup of the keyword "Signal" in all text files and file contents to confirm if there are any Signal-related artifacts within the backup. The search turns 98 hits, including SMS, contact, process activity, contact activity, SQLite databases, Plist, and other artifacts. Although the Signal database and shared media are not included in the iTunes backup, we can still tell a lot from the logical extraction.

In the contact and SMS section, the Signal verification code sending service is extracted from the device contact book, and the SMS containing the one-time

Fig. 3.10 Oxygen forensic detective dashboard for Alice iPhone 8 iTunes backup

verification code is parsed. With the timestamp, investigators can understand whether a Signal messenger is installed on the device (Fig. 3.11).

When searching for the OS artifacts, the Signal app-related process records are found in the iTunes backup. This information can tell the investigator that Signal is installed on the device, and when it is launched. Another important piece of information is that the application ID of Signal in an iOS device is "org.whispersystems.signal", and the process ID is "Signal/org.whispersystems.signal" (Fig. 3.12).

The search also reveals the folder structure of Signal on iOS devices, and the folders found from the iTunes backup are listed in the following screen capture. When looking into the folders, no actual database or other artifacts are found within the folders. However, this information can be preserved as a reference if a full

Fig. 3.11 Signal artifacts in contact book and SMS

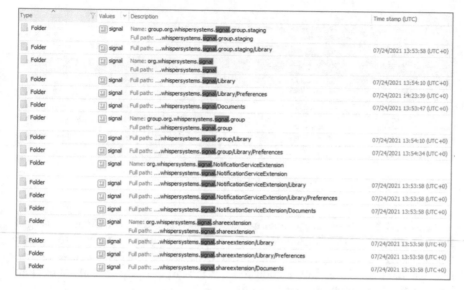

Fig. 3.12 Signal artifacts in contact and process activity

filesystem or physical extraction can be done at a later stage. According to the research conducted by SANS [38], the Signal data can be found in the paths:

- /private/var/mobile/Containers/Data/Application/<APP_GUID>
- /private/var/mobile/Containers/Shared/AppGroup/<APP_GUID>

However, no Signal data is in the above paths.

In the SQLite database "Calendar.sqlite.db", the iCloud account used on the iOS device can be identified. In this test, the email address alice_signaltest@protonmail.com is the iCloud username used on this iPhone 8. Moreover, the two YouTube videos shared by Alice iPhone 8 with Bob iPhone X in Signal chat are captured in the Safari Browser artifacts section.

In the Wireless Connections sections, the Wi-Fi SSID that the iOS device has connected are listed with coordinates, which can be used for further combined searches with other OSINT data to trace the target's activities at a certain point in the timeline.

Another important source of evidence is the Password and Token section. For example, in the iTunes backup, 15 credentials and 188 keychain data are parsed by Oxygen, including the keychain file of WhatsApp, Viber and iCloud. If the cloud account forensics is within the authorization, investigators can adopt these credentials to extract more backup data from the cloud (Fig. 3.13).

The iPhone X is running iOS 14.3 during the tests, and the iPhone 8 is running iOS 14.6, so the difference in the extractions can be identified. Note that fewer artifacts related to Signal are turned in the global search, and no iCloud account information is included in the iTunes extraction.

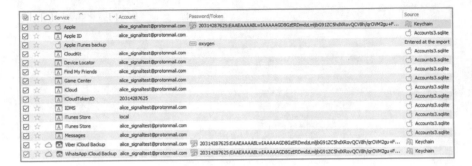

Fig. 3.13 Credentials and keychain artifacts in iTunes backup

In general, an iTunes backup does not provide investigators with Signal messages, shared media, and other related data. Still, it can be used as a reference to indicate whether a Signal messenger is installed on an iOS device or not. Besides, investigators may use the geolocation information to trace the target's activity and adopt cloud investigation credentials and keychain files.

iOS Full Filesystem Extraction with Checkm8

Checkm8 bootrom is the major solution for the forensic acquisition of iOS devices. The open-source iOS jailbreaking tool is embedded as the standard method to create a full filesystem extraction of the iOS device. Oxygen Forensic Detective, Cellebrite UFED, Belkasoft and other commercial mobile forensic manufacturers rely on the update of Checkm8 to support new iOS versions and later Apple devices. According to the official GitHub site, Chechm8 is a permanent uncatchable bootram exploit for iOS devices allowing dumping SecureRom, decrypting keybags for iOS firmware, and demoting devices for JTAG [39]. Another advantage of iOS full filesystem extraction is that no permanent jailbreaking is required. In the test, a Checkm8 full filesystem extraction is conducted on both the iPhone 8 and the iPhone X using Oxygen Forensic Detective. A second Checkm8 extraction is conducted with Cellebrite UFED Touch 2. Although the Signal foundation criticizes Cellebrite for its lack of security mechanism for the mobile data extraction tool, Cellebrite remains one of the most popular and most robust mobile forensic solutions on the market. Therefore, an extraction is conducted for comparative analysis. The main aim of the extractions is to understand what Signal artifacts can be found from an iOS full filesystem extraction, how they can contribute to the investigation and feed the real-time investigative actions and OSINT research.

With the Checkm8 full filesystem extractions, the Signal database and other related data such as user information are extracted. Therefore, differences between the two test scenarios will also be discussed in a comparative approach.

As listed in Table 3.2, the Signal is configured with default settings in the first test scenario, and more secure settings are configured in test scenario 2. It is worth in that some features in which the usage and results are straightforward are not tested, especially, using system contact photos and always relaying calls.

Table 3.2 Signal settings on iPhone 8

		Scenario 1	Scenario 2
Chat settings	Link preview	✓	✗
	Show chats in suggestions	✓	✗
	Use system contact photos		
Notification content	Name and content	Name, content, and actions	No name or content
	Notify when contact joins signal	✓	✗
Privacy settings	Read receipts	✓	✗
	Typing indicators	✓	✗
	Disappearing messages default timer	✗	5 min
	Hide screen in App switcher	✗	✓
	Screen lock	✗	✓
	Show calls in Recents	✓	✗
	Always relay calls	✗	✗
	Sealed sender	✗	✗

The first difference between the iTunes backup and the Checkm8 full filesystem extraction is the size of the extraction file. The iTunes backup of the iPhone 8 is 87.3 MB, and the Checkm8 iOS full filesystem extraction tar file for test scenario 1 is 9.39, and 9.66 GB for test scenario 2. We can easily conclude that full filesystem extraction should be obtained for iOS devices if possible.

Again, we start the comparative analysis from the Oxygen extraction dashboards (Figs. 3.14 and 3.15). From the two figures below, we can view the different numbers of Signal data extracted by Oxygen. The hypothesis is that the disappearing message timer is set to 5 min in test scenario 2. According to Signal official documentation, if enabled on the sender side, the disappearing message timer is counted from the moment when a message is sent. In test scenario 2, a disappearing message is enabled in the Alice iPhone 8, the sender, but not on Bob iPhone X, the receiver. Therefore, all messages disappear 5 min after sending out.

When we look into the Signal section in Oxygen, we can see that the mobile number, the contact (Bob iPhone X), text messages, shared media and logs are extracted as expected. In addition, the mobile number used to register Signal and the user-defined PIN code are extracted. If authorised, this information can be used for the Signal account takeover discussed in the previous section. Also, the Signal contact's profile image, phone number, and username are extracted. This information can be used for Signal account take over and further OSINT investigation, such as image reserve search, name, and phone number reverse search. Compared with the

Fig. 3.14 Oxygen dashboard of iPhone 8 full filesystem extraction—test scenario 1

Fig. 3.15 Oxygen dashboard of iPhone 8 full filesystem extraction—test scenario 2

test scenario two extraction, the author finds that the same Signal user account and contact information are extracted.

Regarding the messages in the test dataset, the contents and the timestamps are extracted except the message "delete for everyone" and view-once media. The deleted messages have only the timestamps and direction (incoming/outgoing), but with empty content; the same for the view-once media. However, for test scenario 2, the messages disappear 5 min after sending out, and the extraction is conducted after the timer is reached. Therefore, the author notices that the disappearing messages have no artifacts left in the extraction, no content, timestamps or message direction (Figs. 3.16 and 3.17).

Another finding is about the URL preview (Fig. 3.18). Signal generates a URL preview by interpreting the URL when a user pauses to type. However, if the user pauses a bit, and then finishes typing the rest of the URL, Signal will interpret a

○	Signal message	07/24/2021 14:08:49 (UTC +0)		Bob Signaltest <+962799298997>	+962799791376
○	Signal message	07/24/2021 14:12:12 (UTC +0)		Bob Signaltest <+962799298997>	+962799791376
○	Signal message	07/24/2021 14:21:45 (UTC +0)		Bob Signaltest <+962799298997>	+962799791376
○	Signal message	07/24/2021 14:23:01 (UTC +0)		Bob Signaltest <+962799298997>	+962799791376
○	Signal message	07/24/2021 14:25:44 (UTC +0)		Bob Signaltest <+962799298997>	+962799791376
○	Signal message	07/24/2021 14:18:03 (UTC +0)		+962799791376	Bob Signaltest <+962799298997>
○	Signal message	07/24/2021 14:18:53 (UTC +0)		+962799791376	Bob Signaltest <+962799298997>
○	Signal message	07/24/2021 14:22:29 (UTC +0)		+962799791376	Bob Signaltest <+962799298997>
○	Signal message	07/24/2021 14:25:16 (UTC +0)		+962799791376	Bob Signaltest <+962799298997>

Fig. 3.16 Signal messages "delete for everyone" and view-once media artifacts—test scenario 1

			Type	Time stamp (UTC)	Description	From	To
☑ ☆	○	○	Signal message	07/31/2021 09:16:06 (UTC +0)	You set disappearing message time: 00:05:00	Bob Signaltest <+962799298997>	+962799791376

Fig. 3.17 Signal message artifacts—test scenario 2

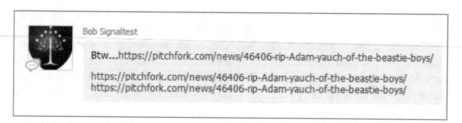

Bob Signaltest

Btw...https://pitchfork.com/news/46406-rip-Adam-yauch-of-the-beastie-boys/

https://pitchfork.com/news/46406-rip-Adam-yauch-of-the-beastie-boys/
https://pitchfork.com/news/46406-rip-Adam-yauch-of-the-beastie-boys/

Fig. 3.18 URL preview artifacts

second URL. For this reason, we can see from the figure below that three URLs are extracted in a single Signal message. It might be not very clear for investigators to distinguish whether they are three different URLs or just one URL with a pause in typing at first. This could be distinguished by reviewing the URLs. It is obvious that the typing was paused or interrupted in our case.

Then, we can investigate where the Signal databases locate. As mentioned before, Signal databases should be in the following paths:

- /private/var/mobile/Containers/Data/Application/<APP_GUID>
- /private/var/mobile/Containers/Shared/AppGroup/<APP_GUID>

However, the APP_GUID may not be obtained by reviewing the arte-facts. Instead, we can search for the keyword Signal in the filesystem. From the search result, we can see the Signal database signal.sqlite locates in the path/private/var/mobile/Containers/Shared/AppGroup/687B747F-FF8F-4C7F-BDA9-1BCE0F7BA064/grdb/signal.sqlite, therefore, the Signal APP_GUID is 687B747F-FF8F-4C7F-BDA9-1BCE0F7BA064. We can then use this information to search other Signal data within the filesystem. From the search, we can understand that the messages and the shared media are stored in the encrypted signal.sqlite. According to the research of Magpol, decrypt the Signal database requires the database key in the keyvalue "GRDBDatabaseCipherKeySpec" is randomly gener-ated when initialising the Database for the first time. The Key and value are then

Fig. 3.19 GRDBDatabaseCipherKeySpec Keyvalue

stored in the iOS keychain. The keyvalue is extracted by the iOS full filesystem extraction and decoded using base64 (Fig. 3.19).

When comparing the two Signal databases from the iPhone 8 of test scenario 1 and scenario 2, the author finds that the sizes of the two databases are the same, both are 1.31 MB. Given that the same testing dataset is used for the two scenarios, it is worth researching if the disappearing messages are saved in the database but hidden from view.

Another important artifact is the group.org.whispersystems.signal.group.plist locates in the path /private/var/mobile/Containers/Shared/AppGroup/APP_GUID /Library/Preferences, which contains the Signal version information and the app configuration data (Fig. 3.20).

Similarly, the Signal settings on the iPhone X are different in test scenario 1 and test scenario 2 (Table 3.3).

Like the iPhone 8, the Signal user and contact information are extracted. The Signal database is the same as mentioned for iPhone 8. Besides the Signal data, a full filesystem provides investigators with other sources of evidence, such as GPS geolocation information that can be used to create a map of the target's activity. The map information can support social media analysis to validate where the target locates at a specific time point.

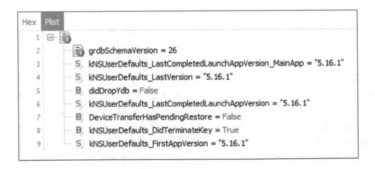

Fig. 3.20 group.org.whispersystems.signal.group.plist

Table 3.3 Signal settings on iPhone X

		Scenario 1	Scenario 2
Chat settings	Link preview	✓	✓
	Show chats in suggestions	✓	✓
	Use system contact photos	✗	✗
Notification content	Name and content	Name, content, and actions	Name, content, and actions
	Notify when contact joins signal	✓	✓
Privacy settings	Read receipts	✓	✓
	Typing indicators	✓	✓
	Disappearing messages default timer	✗	✗
	Hide screen in App switcher	✗	✗
	Screen lock	✗	✗
	Show calls in Recents	✓	✓
	Always relay calls	✗	✗
	Sealed sender	✗	✓

The keychain files and credentials extracted from the iOS keychain can be used for cloud investigation and OSINT investigation (Fig. 3.21).

Regarding the Signal database, the findings on iPhone X also indicate that the size of the signal.sqlite remains the same even if the disappearing message timer is enabled. The author uses Magnet AXIOM which supports manual extraction of GRDBDatabaseCipherKeySpec keyvalue, and then uses the keyvalue to decrypt the imported Signal database. After importing the signal.sqlite extracted from the iPhone X full filesystem extraction created with Oxygen, and copy the GRDBDatabase-CipherKeySpec keyvalue decoded base64 by Oxygen, import them into Magnet AXIOM for processing. Once the processing finishes, we can find the Signal chats

Fig. 3.21 Signal account and contact information—iPhone X

Fig. 3.22 Signal database
path in Cellebrite Physical
Analyzer

under the Communication Artifact type. Although Magnet AXIOM allows users to
decrypt the imported Signal database, it does not provide a view of the decrypted
Signal database signal.sqlite.decrypted. To view the database structure, the author
uses Cellebrite Physical Analyzer.

The Signal database extracted from the iPhone X for test scenario 1 is
located in the path Apple_iPhone X (A1901).zip/root/private/var/mobile/
Containers/Shared/AppGroup/8CF8600C-3A28-49FB-93B1-CCB85FE3FCFC/
grdb/signal.sqlite. Under this path there is another decrypted version Signal
database signal.sqlite.decrypted, which is our target of analysis (Fig. 3.22).

From the file info tab in Cellebrite Physical Analyzer, we know that the size of the
decrypted Signal database is 1351680 Bytes with 4830 rows. In the database view
tab, we can clearly view the Signal iOS database structure. The Signal database is
an SQLite database, and when looking into the sqlite_master table, we know that the
Signal database has 139 database tables in total, and 18 of them are deleted tables
in this case (Fig. 3.23). Not all tables will be covered in this chapter. We will focus
on the database tables containing the messages, user information, timestamps, and
verify if the view-once media read and deleted messages can be found in the Signal
database.

The first table we will explore is the model_OWSUserProfile (Fig. 3.24). In this
database table, the user profile name, the phone number linked to the Signal account,
the uniqueID of the user, the last profile fetch timestamp, the last messaging timestamp, the profile image path, and filename can be found. With the information from
this database table, investigators can conduct OSINT searches on the name, phone
number and profile images. In addition, the uniqueID of users, can be used to query
the database to fetch all user information.

It is also worth noting that Signal iOS uses Unix epoch time to represent timestamps in the database. The time encoding is identified automatically in Cellebrite
Physical Analyzer.

The second database table to explore is indexable_text, where the Signal messages
are stored. In this database table, we can find the message content and the unique
messages ID generated by Signal. The opened view-once media and deleted messages

type ▾	name	tbl_name
table	model_TSGroupMember	model_TSGroupMember
index	index_model_TSPaymentModel_on_isUnread	model_TSPaymentModel
index	index_model_TSPaymentModel_on_mcTransactionData	model_TSPaymentModel
index	index_model_TSPaymentModel_on_mcReceiptData	model_TSPaymentModel
index	index_model_TSPaymentModel_on_mcLedgerBlockIndex	model_TSPaymentModel
index	index_model_TSPaymentModel_on_paymentState	model_TSPaymentModel
index	index_model_TSPaymentModel_on_uniqueId	model_TSPaymentModel
index	sqlite_autoindex_model_TSPaymentModel_1	model_TSPaymentModel
table	model_TSPaymentModel	model_TSPaymentModel
index	index_model_TSInteraction_on_uniqueThreadId_and_erald_and_recordType	model_TSInteraction
index	index_model_TSInteraction_on_uniqueThreadId_and_hasEnded_and_recordType	model_TSInteraction
index	index_model_TSThread_on_isMarkedUnread_and_shouldThreadBeVisible	model_TSThread
index	index_model_TSMention_on_uniqueMessageId_and_uuidString	model_TSMention
index	index_model_TSMention_on_uuidString_and_uniqueThreadId	model_TSMention
index	index_model_TSMention_on_uniqueId	model_TSMention
index	sqlite_autoindex_model_TSMention_1	model_TSMention
table	model_TSMention	model_TSMention
index	index_model_TSInteraction_on_uniqueThreadId_and_attachmentIds	model_TSInteraction
index	index_model_TSInteraction_on_uniqueThreadId_recordType_messageType	model_TSInteraction
index	index_model_TSAttachment_on_uniqueId	model_TSAttachment
index	index_model_TSAttachment_on_uniqueId_and_contentType	model_TSAttachment
index	index_model_OWSReaction_on_uniqueMessageId_and_read	model_OWSReaction
index	index_model_IncomingGroupsV2MessageJob_on_groupId_and_id	model_IncomingGroupsV2MessageJob
index	index_pending_read_receipts_on_threadId	pending_read_receipts
table	pending_read_receipts	pending_read_receipts
index	index_model_ExperienceUpgrade_on_uniqueId	model_ExperienceUpgrade
index	sqlite_autoindex_model_ExperienceUpgrade_1	model_ExperienceUpgrade
table	model_ExperienceUpgrade	model_ExperienceUpgrade
index	index_model_IncomingGroupsV2MessageJob_on_uniqueId	model_IncomingGroupsV2MessageJob

Fig. 3.23 sqlite_master table of signal database

Fig. 3.24 model_OWSUserProfile database table of signal iOS

have a database entry, but the message content is empty. The corresponding timestamps and the direction information of the messages can be found in a different database table index_interactions_on_timestamp_sourceDeviceId (Fig. 3.25).

What are still missing from the Signal database? The data of deleted messages, the shared files, and the view-once media. The scrambled content of deleted messages can be found in the database table indexable_text_fts_data (Fig. 3.26). Although not all message content can be recovered, a large part of the information can be reconstructed.

The shared files are considered attachments in the Signal database, and they can be located in model_TSAttachment. In addition, the message ID of the view-once media and whether they have been opened can be found in the database table

Fig. 3.25 indexable_text database table of signal iOS

Fig. 3.26 indexable_text_fts_data database table of signal iOS

index_interactions_on_view_once. However, the view-once media contents are not located in the Signal database.

The decrypted Signal database of test scenario 2 is also reviewed for comparison. Especially that the disappearing message timer is set as 5 min. The path of the decrypted Signal database is Apple_iPhoneX.zip/root/private/var /mobile/Containers/Shared/AppGroup/FDF7354F-7FF5-485C-8A1C-C6CCF49A 8037/grdb/signal.sqlite/signal.sqlite.decrypted.

When we investigate the database table indexable_text where the message contents are stored, the author finds that the message contents are either empty or replaced with the sender's phone number username (Fig. 3.27).

Another database table to investigate is model_OWSDisappearingMessages Configuration, in which we can find the unique ID of the user who enables the disappearing message timer and the timer in seconds (Fig. 3.28).

id ▼ ▲	collection ▼	uniqueId	▼	ftsIndexableContent	▼
1	TSThread	D710D524-B2CF-438C-B31B-0B419BB26253			
1	TSThread	D710D524-B2CF-438C-B31B-0B419BB26253			
1	TSThread	D710D524-B2CF-438C-B31B-0B419BB26253			
2	SignalRecipient	C55C0D7C-FA8F-4E72-B8D5-A3C022E09E0B			Alice Signaltest
2	SignalRecipient	C55C0D7C-FA8F-4E72-B8D5-A3C022E09E0B			Alice Signaltest
2	SignalRecipient	C55C0D7C-FA8F-4E72-B8D5-A3C022E09E0B			Alice Signaltest
3	TSThread	674E23A2-4455-479C-ADBC-6EBAC857DC51			Alice Signaltest
3	TSThread	674E23A2-4455-479C-ADBC-6EBAC857DC51			Alice Signaltest
3	TSThread	674E23A2-4455-479C-ADBC-6EBAC857DC51			Alice Signaltest
4	TSInteraction	8256CFBB-7162-43D0-8579-2D0AE86FE2E4			
4	TSInteraction	8256CFBB-7162-43D0-8579-2D0AE86FE2E4			
4	TSInteraction	8256CFBB-7162-43D0-8579-2D0AE86FE2E4			
33	TSInteraction	E91F2E83-9731-4A2A-B511-E35987B2CBBB			

indexable_text (4) (9)

Fig. 3.27 indexable_text database table of signal iOS with disappearing message

model_OWSDisappearingMessagesConfiguration (1) (0)

id ▼	recordType ▼	uniqueId	▼	durationSeconds ▼	enabled ▼
1	39	674E23A2-4455-479C-ADBC-6EBAC857DC51		300	1

Fig. 3.28 model_OWSDisappearingMessagesConfiguration

3.5.5.2   Signal Windows

The forensic acquisition and analysis of Signal Windows and Signal macOS are discussed in this section as they are linked to the iOS devices in the tests. Signal client on Windows 10 examiner laptop is linked to the Alice iPhone 8, and Signal client on the macOS Big Sur is linked to the Bob iPhone X. Both Signal desktop clients are not used to send messages but to synchronize the messages sent and received on the mobile devices. The test aims to simulate how investigators can use a linked device to receive real-time Signal messages, shared media, and files.

For test scenario 1, the disappearing messages timer is not enabled. Therefore, most of the messages sent and received between the iOS devices remain on the interfaces of Signal desktop clients. Only the messages deleted from both participants and the view-once media will disappear from the Signal desktop interface. As for test scenario 2, the disappearing message timer is enabled for five minutes on the Alice iPhone 8. A notification displays on the Signal desktop clients to notify users about the disappearing message settings. Of course, investigators can use screen recording software such as FlashBack to record the Signal desktop interface. However, the screen capture cannot store the shared media and files. Another option to monitor the incoming Signal messages is from Windows notification. When the Signal desktop is configured to include name, content, and action in the notifications, investigators can collect the notifications from reconstructing the chat session.

The advantage of Windows 10 system notification is that the notifications remain in the notification pane even when the Signal desktop is shut down. However, notifications still cannot capture the media and files shared. Therefore, the best option is to acquire the desktop device and understand what Signal artifacts can be retrieved.

When the test scenarios are finalized, the Windows 10 examiner laptop is not powered off immediately. Instead, a RAM dump is created with FTK Imager and imported to Magnet AXIOM with Volatility memory analysis framework included for analysis. When reviewing the artifacts parsed and carved by Magnet AXIOM, we find that the images shared within the Signal test chats are captured in Windows RAM images. The Signal profile image of the Alice iPhone 8 and Bob iPhone X is also captured. Moreover, when searching the profile images in Google, the other images are also captured in the Windows 10 RAM image. It is worth noting that the images shared in the deleted Signal messages and sent as view-once media are also found from the Windows 10 RAM dump (Fig. 3.29). As mentioned in the Signal iOS database analysis, the view-once media are not located clearly in the Signal iOS database. With the memory dump of the linked Windows desktop, investigators can find the artifacts on the desktop.

The Web URLs typed in Signal chat as text messages are found in Windows 10 RAM dump and carved by Magnet AXIOM as browser activity. In addition, the native format URLs are listed as potential browser activity, and the corresponding Webpage titles are shown in the Firefox SessionStore artifacts (Fig. 3.30).

When reviewing the Windows LNK files in Windows 10 RAM dump (Fig. 3.31), we find the Signal desktop installation path and the MAC timestamps. Also, the device Mac address of the desktop device network adapter is captured. With this information, investigators can map the devices linked or linked with a target Signal account, which could be used for network forensic analysis.

Another important finding from Windows 10 RAM dump is the process information of the Signal desktop. On Windows 10, Signal.exe launches process 7156, and four other processes share the same parent process 7156. The five processes are

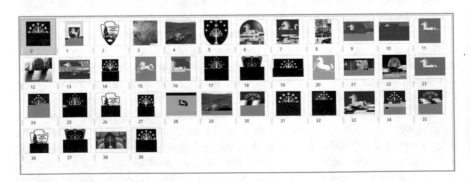

Fig. 3.29 Signal images found from linked Windows 10 RAM dump

Item	Type	Artifact category
https://www.YouTube.com/watch?v=x8mUixkEyJs	Potential Browser Act...	Web Related
https://www.youtube.com/watch?v=fUeJtM1bgGo	Potential Browser Act...	Web Related
Top Cyber Attacks In History | Biggest Cyber Attacks Of All Time | Cyber Security | Simplilearn	Firefox SessionStore...	Web Related
httpsww.y ¿ýĹom/watch?v=fUeJt€bgGo	Firefox SessionStore...	Web Related
Animal Collective - Centipede Hz (Album Trailer)	Firefox SessionStore...	Web Related

Fig. 3.30 URLs sent in signal chat captured in Windows 10 RAM dump

Linked Path	Target File Created Date/Time	Target File Last Modified Da...	Target File Last Accessed Date/Time	MAC Add...
C:\Users\yansh\AppData\Local\Programs\signal-desktop\Signal.exe	7/24/2021 13:17:01	7/21/2021 21:08:08	7/24/2021 13:17:05	60:6D:3C:50:FE:0E
C:\Users\yansh\AppData\Local\Programs\signal-desktop\Signal.exe	7/24/2021 13:17:01	7/21/2021 21:08:08	7/24/2021 13:17:04	60:6D:3C:50:FE:0E

Fig. 3.31 Signal desktop information in Windows 10 RAM dump

Process Name	Process ID	Parent Process ID	Number of Threads	Handles	Session ID	Process Start Date/Time	Location
Signal.exe	7156	9560	29	0	1	7/24/2021 13:54:48	File Offset 4801769600
Signal.exe	2552	7156	13	0	1	7/24/2021 13:54:49	File Offset 9171501248
Signal.exe	9836	7156	7	0	1	7/24/2021 13:54:50	File Offset 9160908928
Signal.exe	22556	7156	40	0	1	7/24/2021 13:54:51	File Offset 11158442176
Signal.exe	18824	7156	7	0	1	7/24/2021 13:54:53	File Offset 9160650944

Fig. 3.32 Memory processes of signal windows

created for the same Signal session and started in a sequential style. The process start times are also captured, but no exit timestamps are linked because the Signal desktop is opened while the RAM is captured. If a deep analysis of the Signal desktop behaviour is needed, investigators can also investigate the dynamically loaded libraries (DLL) list linked with Signal.exe (Fig. 3.32).

Besides the Windows 10 RAM dump, the Windows 10 examiner laptop is also powered off for a post-mortem full disk acquisition. The Windows 10 laptop is booted into a bootable USB thumb drive with CAINE 12 Beta Linux distribution for a live acquisition. A forensic image of EnCase Expert Witness E01 format is created for the 256 GB internal SSD of the laptop. The forensic image is then processed by Magnet AXIOM.

Previously, we mentioned that the Windows 10 system notifications can be a source to extract incoming Signal messages. When reviewing the artifacts within the full disk image in Magnet AXIOM, the author finds that not all incoming messages from Bob SignalTest to Alice SignalTest, the Signal account used on the iPhone 8 and Signal Windows, are captured in the Windows 10 notification. The Windows 10 notifications are stored in the SQLite database wpndatabase.db in the system path in XML format:

<SYSTEM_PARTITION>   \Users\<USER_NAME>\AppData\Local\Microsoft\
Windows\Notifications

Title	Subtext	Received Date/Time	Expiration Date/Time
Bob Signaltest	wait what? leadiing me on for what o___o	7/24/2021 14:01:53	7/27/2021 14:01:53
Bob Signaltest	its a misunderstanding YES LOL who did you hear that from	7/24/2021 14:05:26	7/27/2021 14:05:26
Bob Signaltest	its just a misunderstanding LOL and uh its okay you dont need to feel...	7/24/2021 14:08:51	7/27/2021 14:08:51
Bob Signaltest	Wow That is sth great I have sth too Look at this	7/24/2021 14:12:08	7/27/2021 14:12:08
Bob Signaltest	📷 View-once Photo	7/24/2021 14:12:19	7/27/2021 14:12:19
Bob Signaltest	Btw...https://pitchfork.com/news/46406-rip-Adam-yauch-of-the-beast...	7/24/2021 14:16:32	7/27/2021 14:16:32
Bob Signaltest	Nice~~~~~~	7/24/2021 14:25:38	7/27/2021 14:25:38

Fig. 3.33 Captured incoming signal messages in wpndatabase.db

The database can be viewed from Magnet AXIOM SQLite Viewer (Fig. 3.33). As not all incoming Signal messages are stored in the database, Windows system notification should not be considered the only source of Signal information during an investigation. However, the value of Windows notification is that the message content of deleted messages may be found, in this case, the message from Bob *"its just a misunderstanding LOL..."* is deleted from both iOS devices but is found in the Windows 10 full disk image.

To better understand how Signal stores data in Windows 10 and extract as much information from the application, we will focus on the Signal database now. The Signal folder can be found in the system path: <SYSTEM_PARTITION>Users\<USER_NAME>\AppData\Roaming\Signal\. Another the multiple subfolders, we will focus on two files:

- config.json
- SQLite database db.sqlite in the sql subfolder.

According to the research blog of Judith Myerson published on TechTarget [40], each time Signal desktop opens the database, it stores the encryption key in plaintext to the local configuration file config.json. Just as mentioned in the blog, the content of config.json can be viewed within Magnet AXIOM, and the keyvalue is:

```
31fe07297864bf5413f372c303df10a1b3dcf27e0bba1c8929dd4db
58d5f9f3e
```

Many information security researchers have criticized how Signal stores a database's encryption key, but Signal responded that at-rest encryption is never something that Signal promises to provide. However, how Signal stores its database encryption key remains a risk, as any malicious users with access to the physical machine and the device's administrator access can decrypt the Signal database. In this case, we export sb.sqlite and use the plaintext encryption key to decrypt and open the Signal database in DB Browser (SQLCipher). The decrypted database structure is shown in the screen capture below (Fig. 3.34).

The message content, send, and receive timestamps, type, and other message properties can be found in the messages database table. The timestamps in the Signal desktop database are represented in Unix epoch milliseconds. Therefore, we can convert the timestamps into the human-readable format by writing a simple SQLite query as below:

Fig. 3.34 Decrypted signal desktop database from Windows 10

```
SELECT                    id,conversationId,type,source,body,datetime
(sent_at/1000, 'unixepoch', 'utc'), datetime (received_at/1000,
'unixepoch', 'utc'), datetime (sent_at/1000, 'unixepoch', 'utc'),
hasAttachments,  hasFileAttachments,  hasVisualMediaAttachments
FROM messages;
```

The author noticed that the contents of deleted messages are empty in the Signal database, but there is a database entry recorded in the messages table. Also, Signal uses a Boolean database variable to define if there is an attachment, file attachment, and visual media attachment for each message. The author then finds that the attachments of Signal messages are stored in another subfolder attachments.noindex and categorized as archive files by Magnet AXIOM. However, when exported, the attachments can be opened with the default apps (Fig. 3.35).

3.5.5.3 Signal MacOS

The MacBook Pro running macOS Big Sur has a Signal for Mac installed and linked to Bob iPhone X for both test scenarios. When the tests are finished, the MacBook Pro remains powered on, and a live RAM dump is captured with OSXPMem 2.1, and a Mac memory image in RAW format is created. Similar to Windows 10, not

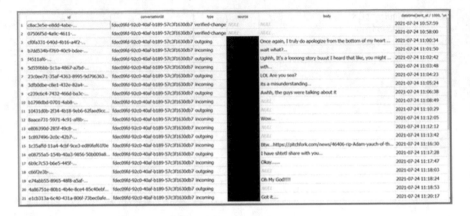

Fig. 3.35 Screen capture of signal database

many forensic findings related to Signal in the memory dump. Only the URLs typed as Signal messages are captured as network browser activity (Fig. 3.36).

Then, the MacBook Pro is powered off for a full disk forensic acquisition. CAINE 12 Beta bootable USB is used to create the forensic image in E01 format of MacBook Pro. The forensic image is then imported into Magnet AXIOM for processing. Before we focus on the Signal database, we first look at the KnowledgeC database. In this database, we can find Signal application focus data. The start and end timestamps record the usage on Signal for Mac. Besides, the KnowledgeC database also records the incoming Signal notifications. However, similar to Windows 10, not all incoming Signal messages are written into the notification database. Note that on both Windows 10 and macOS, a complete data deletion of Signal data from the clients and removing the Signal desktop for a fresh installation does not necessarily remove all traces of Signal usage. The network and application usage artifacts record the usage of Signal.

Item	Type	Artifact category
https://cdn2.signal.org/attachments/Canv9ySjDE4XoIuVcmrc	Potential Browser Activity	Web Related
Top Cyber Attacks In History | Biggest Cyber Attacks Of All Time | Cyber...	Firefox SessionStore Artifacts	Web Related
https://www.youtube.com/watch?v=fUeJtM1bgGo	Potential Browser Activity	Web Related
https://www.youtube.com/watch?v=x8mUixkEyJs	Potential Browser Activity	Web Related

Fig. 3.36 URLs found in Mac memory dump

Fig. 3.37 Signal database messages table in macOS with disappearing message enabled

Then, we will focus on the Signal database in macOS. For comparison purposes, the Signal database for test scenario 2 with a disappearing message timer set is extracted for a detailed analysis here. The Signal data storage locates in the system path.

<MAC_SYSTEM_PARTITION>\Users\<USER_NAME>\Library\Application Support\Signal\

Signal for Mac has the same folder structure as Signal Windows, which facilitates cross-platform development and forensic exanimation. Same as in Windows 10, we can locate the Signal database in sql subfolder, and we can also extract the SQLite database encryption key in plaintext from file config.json. The same procedure used to decrypt the Signal database in Windows is used for macOS.

In the Signal database, not all the messages sent and received in the test dataset are listed in the messages database table (Fig. 3.37). The disappearing message timer is enabled on the iPhone 8 side, and the full disk acquisition is created after the timer expired, therefore, the empty message contents can be expected. However, unlike the messages deleted using the Signal feature "delete for everyone", which leaves a database entry with empty content, Signal seems to remove the disappearing messages randomly from the database. As a result, some messages have no database entries, while the others can be retrieved. The detailed mechanism of how Signal is used to "sanitize" the database for disappearing messages remains further investigated.

3.5.5.4 Signal Android

Due to the lack of physical Android devices, the author uses the BlueStacks Android emulator on a Windows 10 examiner laptop. BlueStacks provides a native user experience up to Android 10, and the emulator provides various built-in virtual device profiles without manual configuration. In the tests of this chapter, a Samsung Galaxy S10 and a Samsung Galaxy S20 Ultra virtual devices are created in BlueStacks using the pre-defined profiles. The data of BlueStacks Windows can be found in the system path: C:\ProgramData\BlueStacks_nxt.

Each virtual device created in BlueStacks has a separate folder with the naming convention ANDROID_ALIAS_ID No, such as "Nougat64_1". The Android system profile used in the tests is Android Nougat 64-bit. BlueStacks asks users to create a virtual device as the master device. In the folder of the master device, "Nougat64" in our case, we have the folder structure as shown in the screen capture below. From the master device folder, we can see that BlueStacks uses VDI virtual disk format to store data. Based on the master device, more virtual devices with the same configurations can be created by cloning the master device. In the non-master device folder, only the DATA.VDI can be found. The DATA.VDI is the main file that we will focus, on and where the Signal data locates (Fig. 3.38).

Fig. 3.38 BlueStacks master device folder content

Like the tests with iOS devices, the Signal Android are configured with customized settings for test scenarios with a disappearing message timer set as five minutes (Table 3.4).

We firstly have a look at the DATA.VDI of Samsung S10. In the password and token artifacts type list, the Gmail account used to register Google Play Store is stored in the SQLite database in the path <ANDROID_PARTITION>\system_cd\0\accounts_ce.db. The Gmail account properties can be found in the database and the tokens for accessing other Google services. In the same path of accounts-ce.db, the recent system task information is stored in XML format. For example, the file 14_task.xml is the task of Signal Android. The most important information that can be extracted from the XML file is the timestamps when Signal is lastly launched and moved. The token can be used later for extracting cloud data from Google services if authorized.

The images shared in the Signal testing chat can be found from the unallocated space of the virtual Android Samsung device. Also, another copy of the images is found in the cache of the BlueStacks file manager app, which is an app to simulate the Android file manager on a physical device. Therefore, it is worth verifying the finding on the physical Android mobile. Moreover, in the folder <ANDROID_PARTITION>\media\0\DCIM\SharedFolder\ (Fig. 3.39), all image files and other file attachments shared in the test Signal chat are found. However, the finding does not mean Signal Android stores the files, but because the files are first uploaded to the virtual Android devices, and then shared within Signal. The inspiration of this finding is that many forensic artifacts may be stored in the file system, so investigators should not only focus on the Signal folder but should conduct a full analysis of the forensic image acquired.

Table 3.4 Signal settings on virtual Android devices

		Alice Samsung S10	Bob Samsung S20 Ultra
Chat settings	Link preview	✗	✓
	Show chats in suggestions	✗	✓
	Use system contact photos	✗	✗
	Use system emoji	✗	✓
	Chat backup	✗	✓
Notification content	Name and content	No name of content	Show name and message
	Notify when contact joins signal	✗	✓
	Repeat alerts	Never	Never
	Priority	High	High
	Calls notification	✓	✓
	Read receipts	✗	✓
	Typing indicators	✗	✓
Privacy settings	Disappearing messages default timer	✗	✗
	Screen security	✓	✗
	Screen lock (not support)	✗	✗
	Incognito keyboard	✓	✗
	Always relay calls	✓	✗
	Signal messages and calls	✓	✓
	Sealed sender	✗	✓

Name	Type	File...	Size...	Created	Accessed	Modified
17-f0028439_4n6rodeo3_fix_copy.zip	Image	.zip	147,150	7/30/2021 7:43:20 PM	7/31/2021 1:57:38 PM	7/30/2021 7:43:20
24-f0045964_Statements_of_Financial_Condition...	File	.doc	71,680	7/30/2021 7:43:20 PM	7/31/2021 1:56:33 PM	7/30/2021 7:43:20
37-image_0.jpg	File	.jpg	144,695	7/30/2021 7:43:20 PM	7/30/2021 7:43:20 PM	7/30/2021 7:43:20
6-f0003868.jpg	File	.jpg	287,186	7/30/2021 7:43:20 PM	7/30/2021 7:43:20 PM	7/30/2021 7:43:21
Edward Sharpe & The Magnetic Zeros - Man on...	File	.mp4	10,109,144	7/30/2021 7:43:21 PM	7/30/2021 7:43:21 PM	7/30/2021 7:43:21
LOL.PNG	File	.PNG	805	7/30/2021 7:43:26 PM	7/30/2021 7:43:26 PM	7/30/2021 7:43:26

Fig. 3.39 Android artifacts of files shared in signal

The runtime-permissions.xml file in the path <ANDROID_PARTITION>\system \users\0\ keeps records of the system permissions required by each application installed. The approval result is represented with a Boolean variable, 1 is approved, and 0 is rejected. In the XML, we can find the permissions that Signal Android asks for. They are listed in the screenshot below. For example, Signal asks permission to read and write to the device contact book, get account information, read phone state

S. Yan et al.

and external storage, read send SMS and MMS, initiate, and receive calls, camera, and audio records.

The author notices that Signal Android is named org.thoughtcrime.securesms in the filesystem. Therefore, investigators should search for this alias instead of Signal within the Android filesystem. The Signal data folder locates in the path:

<ANDROID_PARTITION>\data\org\thoughtcrime.securesms\

Inside this path, several subfolders can be found, as shown in the screen capture below.

The Signal SQLite database signal.db locates in the subfolder databases. The Signal Android database is encrypted with a different encryption mechanism than in the iOS system; the AES_GCM mode encryption requires keyvalues from the following files to decrypt the Signal database.

- \misc\keystore\user_0\<SIGNAL_APP_ID>_USRSKEY_SignalSecret

The value of hex offset 2D to 3C in the file contains the AES_GCM encryption key;

- \data\org.thoughtcrime.securesms\shared_prefs\org.thoughtcrime. securesms_preferencesxml

The keyvalues of key "pref_database_encrypted_secret" are required.

- The first keyvalue "data" contains the AES-GCM ciphertext, and the last 32 characters consist of the auth tag.
- The second keyvalue, "iv" is the initialization vector.

The encrypted Signal databases are exported from the two virtual Android devices, and the author tries to decrypt the databases with the above keyvalues in Magnet AXIOM; however, the tests are failed. Although Magnet AXIOM does not throw any error message, it returns no decrypted result (Fig. 3.40).

Although the encrypted database cannot be decrypted, some Signal artifacts are still parsed by Magnet AXIOM. The user information and the contact list are parsed.

3.5.5.5 Signal Linux

Signal Linux is installed in the virtual machine hosted in VMWare Workstation Pro 16 on a Windows 10 examiner laptop. Ubuntu 20.04.2 LTS 64-bit is used as the testing operating system. Signal Linux is installed in the Ubuntu virtual environment and linked to the virtual Samsung S10 for the test. When the test is finished, the Ubuntu virtual machine remains powered on, and a snapshot in VMEM format is created. Then, the Ubuntu virtual machine is powered off, and the VMDK virtual disk image is imported into Magnet AXIOM together with the snapshot image for processing (Fig. 3.41).

The VMEM snapshot is like a RAM capture on Windows, in which the network activities of Signal are found, as well as the thumbnail of the URL previewed in the Signal chat.

Fig. 3.40 Decrypt signal android database in Magnet AXIOM

Item	Type	Artifact ca...
https://textsecure-service.whispersystems.org/v1/messages/8bb1252c-716d-4e0a-8a5d-fa7954a3e883	Potential Browser Activity	Web Related
https://api.directory.signal.org	Potential Browser Activity	Web Related
https://cdn2.signal.org	Potential Browser Activity	Web Related
https://cdn.signal.org	Potential Browser Activity	Web Related
https://sfu.voip.signal.org/	Potential Browser Activity	Web Related
https://storage.signal.org	Potential Browser Activity	Web Related
http://contentproxy.signal.org:443	Potential Browser Activity	Web Related
https://signal.org/download	Potential Browser Activity	Web Related
https://cdn2.signal.org/attachments/O62BHcxGDU1fVDETDwpo	Potential Browser Activity	Web Related
https://textsecure-service.whispersystems.org+A	Potential Browser Activity	Web Related
https://textsecure-service-staging.whispersystems.org	Potential Browser Activity	Web Related
https://storage-staging.signal.org	Potential Browser Activity	Web Related
https://api-staging.directory.signal.org	Potential Browser Activity	Web Related
https://updates2.signal.org/desktop	Potential Browser Activity	Web Related
https://textsecure-service.whispersystems.org	Potential Browser Activity	Web Related
https://cdn-staging.signal.org	Potential Browser Activity	Web Related
https://cdn2-staging.signal.org	Potential Browser Activity	Web Related
https://github.com/signalapp/Signal-Desktop.git	Potential Browser Activity	Web Related
https://github.com/signalapp/better-sqlite3#2fa02d2484e9f9a10df5ac7ea4617fb2dff30006	Potential Browser Activity	Web Related
https://github.com/signalapp/signal-ringrtc-node.git#868f7ecb699b984171b5ad02f9b043bfa55ad804	Potential Browser Activity	Web Related
https://github.com/signalapp/signal-zkgroup-node.git#3bb62fa44dc69560436a8c946ea48630f3230ed3	Potential Browser Activity	Web Related
Animal Collective - Centipede Hz (Album Trailer)	Firefox SessionStore Artifacts	Web Related

Fig. 3.41 Signal network activities in Ubuntu VM snapshot

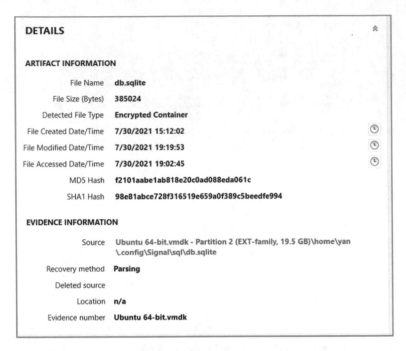

Fig. 3.42 File details of encrypted signal database in Ubuntu

In the virtual disk image VMDK, we can find the Signal database listed in the list of the encrypted files of Magnet AXIOM, which uses the same encryption mechanism as on Windows and macOS. Therefore, we can export the Signal database and decrypt it with the keyvalue from config.json. As expected, the Signal database in Ubuntu has the same database entries as in Windows 10, this is because Signal desktop is developed as a cross-platform application that one-time development can deploy to different platforms (Fig. 3.42).

3.5.5.6 Sharing Signal Data with Investigators

Both criminal investigation and workplace investigation have a clear standard of evidence acceptance. Screen captures from WhatsApp, Skype and other instant messaging applications are one of the main approaches for witnesses, victims and subjects to share information with investigators. However, screen captures are hard to distinguish whether they are authentic as they are usually stored in formats like JPG, PNG or TIFF to send through an insecure channel. Especially, free online resources such as fakechatapp.com, and fakewhats.com allow non-tech savvy users to create fake WhatsApp chats in several minutes. In WhatsApp, exporting chat into a compressed file of text messages and media files provides an alternative solution; however, the compressed file can be extracted. The text file can also be edited in any text editor. For instant-messaging applications with local or cloud data backup functions, a data backup should be created and protected with passwords. Then the package and passwords should be shared with investigators separately. However, Signal only supports data backup on the Android platform, and the backups are encrypted.

The author proposes an alternative solution to share Signal data with investigators. Signal can be installed on an examiner's laptop, and then the generated QR code is shared with the witness to scan. Once scanned, the examiner's laptop becomes a linked device to the witness's mobile handset. Then, investigators can store the synchronized Signal chats via screen recording, screen captures, and forensic acquisition. Once the evidence collection finishes, the witness can remove the examiner's laptop from the Signal linked device list. However, user privacy should be protected, and a clear authorization with a scope well-defined should be reviewed and signed by the witness in advance.

3.6   Discussion

3.6.1   General Process to Handle Signal

With the shifting of the social gravity from personal interaction to online chatting, instant messaging applications have become copies of our lives. Digital forensic investigations on instant messaging applications start with forensically in the mobile device data and reviewing the shared messages and media. However, the development of applications like Signal is driving the change of mindset of mobile forensics to integrated digital forensics. Delete messages from all chat participants, timed disappearing messages, and view-once media pose the same question to investigators and digital forensic practitioners, what to do if post-mortem filesystem/physical acquisition cannot recover deleted data? Also, linked devices of mobile and desktop devices make the physical device seizure more difficult. The target Signal account can remain active with data exchange with only one linked device not seized. Additionally, suppose, the Signal registration lock is not enabled. In that case, the target of an investigation can easily register the same Signal account on a new mobile handset, which makes the seizure non-valid. Not to mention that the target can unlink all devices and delete Signal clients and the account from physical devices, which can lead the investigation to a dead end. These questions naturally lead to further thinking about the importance of extending Signal investigations from mobile examination centred or physical device forensic examination centred mode to an OSINT-based and real-time data flow mode investigation that does not necessarily involve physical devices. This framework relies on the power of OSINT and Signal built-in features to establish a social graph of the target and explore the possibility of monitoring the real-time data exchange within Signal. The physical devices, if seized and acquired, should be considered as a new source of lead that feeds into the real-time and OSINT-based investigation. Besides an in-depth analysis of the Signal database, the contacts, geolocations, timeline information and multimedia artifacts extracted from the physical devices would be input to an iteration of the real-time and OSINT-based investigation.

The next natural step is to combine the segments of thinking into a formal framework that streamlines the entire life cycle of the digital forensic investigation of Signal. Investigators and digital forensic practitioners can refer to the framework for their investigation preparation, decision-making, problem-solving during an investigation, and validating their outcome. Compared with other frameworks for instant-messaging investigation, such as IDFIF proposed by Bery Actoriano and Imam Riadi in their research on WhatsApp Web, the framework proposed in this chapter is specifically tuned to include the unique features of Signal and how to leverage these features to establish a real-time data flow and to facilitate the OSINT investigation. As of the writing time, the framework proposed in this chapter is the first digital forensic investigation framework proposed to the author's best knowledge.

Moreover, this framework is also a reference for future research on Signal. It can also be extended based on separate in-depth research of each step within the framework, such as the mechanism of linking devices to a Signal account. Can this mechanism be bypassed or cracked? Finally, researchers can use the framework in this chapter as an assessment reference to evaluate how their research can fit into the life cycle of a digital forensic investigation and contribute to achieving the investigation requirements.

3.6.2   Signal Data and Database

Signal is outstanding because of its top-level security and privacy protection compared with other market competitors. As part of this mechanism, Signal databases are encrypted on all supported platforms on the mobile and desktop. Although the findings in this section may be different on the different device makes and models, the author finds that the Signal Android has the best database encryption mechanism compared with Signal on other platforms. The Signal database on an Android mobile device is encrypted with a mechanism that requires three different keyvalues to decrypt the SQLite database. Due to the limitation of physical devices, the tests on Signal Android are conducted with virtual Android devices. The author tries to export the keyvalues and the SQLite database and decrypt it in Magnet AXIOM, however, no decrypted results were identified from the result. On the other hand, the Signal database encryption key can be easily found in the iOS keychain file. The author successfully exports the Signal iOS database and decrypts it with the keyvalue in DB Browser SQLCipher. In the decrypted Signal iOS database, user information, including name, telephone number, last activity timestamps and profile description is found. Regarding the deleted messages via disappearing message and delete for everyone, the author finds that not all deleted message contents are removed from the Signal iOS database. For some deleted messages, the message content is empty, while all other message properties, including the timestamps can be identified. Other messages have contents in a disordered format or are scrambled with non-readable symbols. The author also finds that some deleted messages whose contents remain untampered in the Signal iOS database, and the full message content can be recovered. Similar findings are noted from Signal desktop for Windows, macOS, and Ubuntu. From a visual examination, it seems that Signal removes deleted message contents from the SQLite database in a random mechanism, which may leave certain message contents unchanged. This direction could be an option for further in-depth research to understand how Signal removes deleted message contents from the database.

When comparing the findings with previous research done by Judge, her research focused on whether different forensic software can parse Signal messages and carve deleted messages from Android and iOS devices. An important difference between the two research is the improvement of mobile acquisition technology. In Judge's research, a full filesystem extraction of iOS devices was unavailable, so not many Signal artifacts could be extracted from the logical extraction. In this chapter, the author can create full filesystem extractions with different mobile forensic software, leading to the successful decryption of the Signal database. On the Android platform, both pieces of research prove that the logical extraction of Android devices does not reflect the Signal database, and the Signal messages were not extracted either. Another improvement of this chapter is the database analysis. The author walks through the Signal database decryption process in this chapter and discusses Signal's encryption mechanisms on both iOS and Android platforms, which was not covered in Judge's research. Also, the author analyses the digital artifacts in the Signal database manually rather than relying on the software parsing functionality, which reveals more details about how Signal data is stored in the database, and what can be found for the deleted messages.

This chapter also covers the desktop Signal clients for Windows, macOS, and Ubuntu. This chapter proves that Signal desktop is developed in a multi-platform style that the same database structure, folder structure and encryption mechanism are re-used in the three desktop platforms. Like iOS, the encryption key can be found easily from the Signal config JSON file in plaintext and can be used to decrypt the SQLite database manually. The research findings also reveal that Signal media can be found from live memory acquisition from linked desktop devices. The findings also prove the importance of including Signal desktop when investigating Signal.

3.6.3   Signal Investigation Without Physical Devices

One of the core findings of this chapter is to explore the possibility of investigating Signal with an OSINT-based and real-time data flow-based mode. In this chapter, the author finds a viable approach to taking over a Signal account with the registration lock disabled. With the method, investigators can register the target Signal account in an examiner's mobile handset and continue data exchange, recording all user and contact information from the taken Signal account. Also, the author proves that a real-time Signal data flow can be established by linking an examiner desktop or adopting the linked desktop device from the target to monitor the real-time data exchange of the Signal account under investigation. In addition, the various options to add new members to a Signal group offer a possibility for investigators to join the same Signal group as the target, monitor the group activities, and collect information about all group members. Of course, the extension of Signal investigation from a physical device-based, lab forensic examination centred model to an interactive, real-time investigation comes with high risks to alert the target of the investigation; therefore, these activities must be conducted with extra caution and plan in advance.

Meanwhile, the author explores Signal from an OSINT-based approach to fully use the built-in Signal features in investigations. Although Signal has fewer social network-style features than WhatsApp, Viber, and other instant messaging applications, it allows users to search Signal users with telephone numbers and view basic Signal user profiles. The read receipts and typing indicators can also be used in timeline construction to map the target's online activities. If combined with the real-time data investigation and digital forensic examination findings, the geolocation, multimedia, contact information, and OS artifacts can be used as new leads for OSINT investigation.

3.7   Summary

This chapter introduces a new operational framework for digital forensic investigation of Signal instant messenger. The framework covers different forensic methods to handle Signal messenger data with or without the physical devices. Also, the chapter discussed how major forensic acquisition techniques can assist investigators in Signal investigations and their limitations. The chapter emphasizes the importance of real-time data capture and OSINT investigation when dealing with Signal and other instant messaging applications. The framework proves that investigators can have different choices when handling Signal messenger. The post-mortem forensic analysis of Signal can be considered valuable source to drive the OSINT investigation in an iterative model. The chapter proposes a new integrated digital forensic investigation framework for handling Signal secure messenger. The framework includes the entire life cycle of a digital forensic investigation, from the trigger to the final report. The framework includes an iterative process for situations with or without physical data carriers seized and acquired. Also, the chapter discusses the possible approaches for investigators to establish a real-time Signal data flow, and the data captured can be used for OSINT investigation. Signal features are tested with an author-defined dataset within different scenarios, and the test results are useful for investigative actions without the physical devices. In addition, the chapter also covers the digital forensic acquisition and analysis process for Signal supported platforms. Different digital forensic acquisition modes are discussed, and Signal artifacts can be found from the extractions. The chapter describes the process to decrypt Signal databases from different platforms and discusses what artifacts can be found and when to find the artifacts. The framework proposed in this chapter can easily be extended or adjusted for other similar instant messaging applications. Moreover, the framework aims to provide operational level guidance so that investigators and digital practitioners can use the framework directly in their daily investigation activities.

References

1. Signal Messenger. (n.d.). *Technology preview: Sealed sender for signal.* [online] Available at: https://signal.org/blog/sealed-sender/. Accessed December 28, 2021.
2. Statt, N. (2021). *Signal sees surge in new signups after boost from Elon Musk and WhatsApp controversy.* [online] The Verge. Available at: https://www.theverge.com/2021/1/7/22218989/signal-new-signups-whatsapp-facebook-privacy-controversy-elon-musk
3. TechCrunch. (n.d.). *Signal's Brian Acton talks about exploding growth, monetization and WhatsApp data-sharing outrage.* [online] Available at: https://techcrunch.com/2021/01/12/signal-brian-acton-talks-about-exploding-growth-monetization-and-whatsapp-data-sharing-outrage/
4. Pocket-lint. (2021). *WhatsApp terms and conditions update: What you need to do and why.* [online] www.pocket-lint.com. Available at: https://www.pocket-lint.com/apps/news/whatsapp/156398-whatsapp-terms-and-conditions-update-what-you-need-to-do-and-why. Accessed December 3, 2021.
5. www.documentcloud.org. (n.d.). *DocumentCloud.* [online] Available at: https://www.documentcloud.org/documents/3723701-Ron-Wyden-letter-on-Signal-encrypted-messaging.html. Accessed December 3, 2021.
6. Whittaker, Z. (n.d.). *In encryption push, Senate staff can now use signal for secure messaging.* [online] ZDNet. Available at: https://www.zdnet.com/article/in-encryption-push-senate-approves-signal-for-encrypted-messaging/. Accessed December 3, 2021.
7. Staff, R. (2020). *U.N. says officials barred from using WhatsApp since June 2019 over security.* Reuters. [online] 23 January. Available at: https://www.reuters.com/article/us-un-whatsapp-idUSKBN1ZM32P. Accessed December 3, 2021.
8. Unhcr.org. (2021). *UNHCR emergency handbook.* [online] Available at: https://emergency.unhcr.org/entry/190105/instant-messaging-on-unhcr-systems. Accessed December 3, 2021.
9. Porter, J. (2020). *Signal becomes European Commission's messaging app of choice in security clampdown.* [online] The Verge. Available at: https://www.theverge.com/2020/2/24/21150918/european-commission-signal-encrypted-messaging. Accessed December 3, 2021.
10. Wikipedia Contributors. (2019). *Signal (software).* [online] Wikipedia. Available at: https://en.wikipedia.org/wiki/Signal_(software).
11. Wired. (n.d.). *How to use signal encrypted messaging.* [online] Available at: https://www.wired.com/story/signal-tips-private-messaging-encryption/
12. Signal Support. (n.d.). *Set and manage disappearing messages.* [online] Available at: https://support.signal.org/hc/en-us/articles/360007320771-Set-and-manage-disappearing-messages. December 4, 2021.
13. Anon. (n.d.). *WhatsApp forensic artifacts: Chats aren't being deleted—Zdziarski's blog of things.* [online] Available at: https://www.zdziarski.com/blog/?p=6143. Accessed July 2, 2021.
14. Security Affairs. (2018). *Signal disappearing messages can be recovered by the macOS client.* [online] Available at: https://securityaffairs.co/wordpress/72315/security/signal-disappearing-messages.html. Accessed December 2, 2021.
15. Signal Support. (n.d.). *Delete for everyone.* [online] Available at: https://support.signal.org/hc/en-us/articles/360050426432-Delete-for-everyone. Accessed December 4, 2021.
16. WhatsApp.com. (n.d.). *WhatsApp help center—How to delete messages.* [online] Available at: https://faq.whatsapp.com/android/chats/how-to-delete-messages/?lang=en. Accessed December 4, 2021.
17. Jain, R. (n.d.). *Here's how you can bypass WhatsApp's "Delete for Everyone" feature.* [online] TechRadar. Available at: https://www.techradar.com/news/heres-how-you-can-bypass-whatsapps-delete-for-everyone-feature. Accessed December 4, 2021.
18. Signal Messenger. (n.d.). *Sometimes once is better than a lifetime.* [online] Available at: https://signal.org/blog/view-once/. Accessed December 4, 2021.
19. Signal Support. (n.d.). *View-once Media.* [online] Available at: https://support.signal.org/hc/en-us/articles/360038443071-View-once-Media. Accessed December 4, 2021.

20. WABetaInfo. (2021). *WhatsApp is rolling out photos and videos set to view once!* [online] WABetaInfo. Available at: https://wabetainfo.com/whatsapp-is-rolling-out-photos-and-videos-set-to-view-once/. Accessed December 4, 2021.
21. Signal Support. (n.d.). *Mark as unread.* [online] Available at: https://support.signal.org/hc/en-us/articles/360049649871-Mark-as-Unread. Accessed December 4, 2021.
22. Signal Support. (n.d.). *Backup and restore messages.* [online] Available at: https://support.signal.org/hc/en-us/articles/360007059752-Backup-and-Restore-Messages. Accessed December 5, 2021.
23. ElcomSoft blog. (2019). *How to extract and decrypt signal conversation history from the iPhone.* [online] Available at: https://blog.elcomsoft.com/2019/08/how-to-extract-and-decrypt-signal-conversation-history-from-the-iphone/.
24. Alissa, K., Almubairik, N. A., Alsaleem, L., Alotaibi, D., Aldakheel, M., Alqhtani, S., Saqib, N., Brahimi, S., & Alshahrani, M. (2019). A comparative study of WhatsApp forensics tools. *SN Applied Sciences, 1*(11), 1–10.
25. Cents, R., & Le-Khac, N.-A. (2020). Towards a new approach to identify WhatsApp messages. In *19th IEEE International Conference on Trust, Security and Privacy in Computing and Communications (IEEE TrustCom-20)*, Guangzhou, China, December, 2020.
26. Marshall, A. M. (2018). WhatsApp server-side media persistence. *Digital investigation, 25*, 114–115.
27. Choi, J., Park, J., & Kim, H. (2017). *Forensic analysis of the backup database file in KakaoTalk messenger* (p. 156). IEEE.
28. Ichsan, A. N., & Riadi, I. (2021). Mobile forensic on Android-based IMO messenger services using Digital Forensic Research Workshop (DFRWS) Method. *International Journal of Computer Applications, 174*(18), 34–40.
29. Karpisek, F., Baggili, I., & Breitinger, F. (2015). WhatsApp network forensics: Decrypting and understanding the WhatsApp call signaling messages. *Digital Investigation, 15*, 110–118.
30. Conti, M., & Dargahi, T. (2017). *Forensics analysis of Android mobile VoIP Apps.*
31. Al-Rawashdeh, A. M., Al-Sharif, Z. A., Al-Saleh, M. I., & Shatnawi, A. S. (2020). *A post-mortem forensic approach for the Kik Messenger on Android* (p. 079). IEEE.
32. Kukuh, M., Riadi, I., & Prayudi, Y. (2018). Forensics acquisition and analysis method of IMO messenger. *International Journal of Computer Applications, 179*, 9–14. https://doi.org/10.5120/ijca2018917222
33. Agrawal, V., & Tapaswi, S. (2019). Forensic analysis of Google Allo messenger on Android platform. *Information and Computer Security, 27*(1), 62–80.
34. Riadi, I., Umar, R., & Firdonsyah, A. (2018). Forensic tools performance analysis on Android-based blackberry messenger using NIST measurements. *International Journal of Electrical and Computer Engineering, 8*(5), 3991–4003.
35. Mahmod, R., & Yusoff, M. (2017). *Forensic investigation of social media and instant messaging services in Firefox OS: Facebook, Twitter, Google+, Telegram, OpenWapp, and Line as Case Studies.*
36. Judge, S. M. (2018). *Mobile forensics: Analysis of the messaging application Signal.* ProQuest Chapters Publishing.
37. Wijnberg, D., & Le-Khac, N.-A. (2021). Identifying interception possibilities for WhatsApp communication. *Forensic Science International: Digital Investigation, 38*(Supplement), 301132. https://doi.org/10.1016/j.fsidi.2021.301132.
38. Egnyte. (n.d.). 300–160.pdf on Egnyte. (Online). Available at: https://sansorg.egnyte.com/dl/whJmm24X57/?. Accessed August 8, 2022.
39. GitHub. (n.d.). GitHub—axi0mX/ipwndfu: open-source jailbreaking tool for many iOS devices. (Online). Available at: https://github.com/axi0mX/ipwndfu. Accessed August 8, 2022.
40. SearchSecurity. (n.d.). How did Signal Desktop expose plaintext passwords? (Online). Available at: https://searchsecurity.techtarget.com/answer/How-did-Signal-Desktop-expose-plaintext-passwords. Accessed August 10, 2022.

Chapter 4
Forensic Analysis of the qTox Messenger Databases

Daniel Meier, Kim-Kwang Raymond Choo⊙, and Nhien-An Le-Khac⊙

4.1 Introduction

Instant messenger (IM) are software tools to exchange messages in real-time over the internet between two or more users. Today, the most popular instant messenger clients like WhatsApp, Facebook Messenger, Threema, Telegram etc. are proprietary software that is owned and developed by large companies. On the other hand, there are also a lot of open-source projects like Jabber based clients, Matrix protocol or the Tox protocol with different clients like Riot [1], qTox [2]. Most people use well known messenger like WhatsApp [3], Telegram and Threema. The spread of these messengers makes it easy for people to connect with each other. Only a small part of people use less known messenger like qTox or Jabber based clients. They often appreciate the independence from large companies and trust in open source software.

In the past couple of years, the amount of cases regarding cybercrimes has risen sharply. We often see the extensive use of End-to-End Encryption (E2EE) IM applications like Telegram, WhatsApp, Signal or Jabber based clients [3–5]. We also noticed the use of qTox messenger over the Tox protocol for communication between the offenders [6, 7]. Because of the implemented encryption algorithm, most digital forensic investigators were not able to read the user database and therefore not able to get the qTox communication of the suspects. They tend to lost important evidence material due to this fact.

The story of Tox began in the year 2013. The idea was to build an unofficial successor of Skype, E2EE and that ran without requiring the use of central servers. The developers of Tox wanted to build a free and easy to use messenger for a wide range of platforms, supporting voice and video chats, file transfer and desktop sharing capabilities and they wanted it to be decoupled from the frontend respectively user-interface. By now the Tox protocol is an open source project at github.com/TokTok (https://github.com/TokTok/c-toxcore) and a lot of clients for different platforms have been released. All clients are using the same core code of the Tox protocol. qTox is the most popular client for the Tox protocol and it is available for Windows, MacOS,

Linux and BSD. qTox is an open source project, developed in C ++ by different users and managed on the platform github (https://github.com/qTox/qTox/). In the current version of writing this chapter (version 1.16.3) the following features are supported and relevant for a forensic consideration: Chats, Group Chats, File Transfers, Audio- and Video Calls. Users are able to install qTox without registration and without specifying an address, e-mail or the real name.

On the forensic investigation point of view, for most of the common IM applications there are software solutions to reconstruct metadata and communication. Or literature is available that describes a way and methods to reconstruct and/or decrypt such information. However, in the case of qTox/Tox it looks different. Existing well known digital forensic software like Magnet Internet Evidence Finder [8], Magnet Axiom [9] or X-Ways Forensics [10] is not able to detect and extract artifacts of the qTox messenger. Moreover, to the best of our knowledge, literature that describes the possibilities and forensic reconstruction of qTox data and communication has not appeared yet. Hence, this paper aims at investigating whether any forensic artifacts of qTox can be recovered from a specific desktop environment with two main research questions: Is it possible to find forensic artifacts of qTox on a system and what kind of artifacts can be forensically discovered when using qTox messenger? And if there are encrypted artifacts, is it possible to decrypt such data? The main contribution of this chapter can be listed as follows:

- Reconstruct artifacts from the hard drive and the memory of the system.
- A solution to decrypt the qTox sqlite databases.
- Experiments show that forensic artifacts can be recovered from the qTox Client in different desktop environments such as Linux, Windows.

Hence, qTox forensic is interesting for law enforcement agencies. qTox is not widely used like WhatsApp or Telegram, but it is very common in the field of sexual abusers and the child exploitation community. Examining contacts and conversations is important for a forensic investigator since they might give important clues to the case.

4.2 Background Concepts

4.2.1 QTox Client

The story of Tox began in the year 2013. Two years after the software giant Microsoft took over the then leading software for voice and video telephony Skype, a group of people in a 4chan (http://rbt.asia/g/thread/S34778013#p34778939) subchannel about technology discussed the privacy of Skype and the cooperation of Microsoft with the NSA. The idea was to build an unofficial successor of Skype, end-to-end encrypted and that ran without requiring the use of central servers. The developers of Tox wanted to build a free and easy to use messenger for a wide range of platforms, supporting voice- and video chats, file transfer and desktop sharing capabilities and they wanted it to be decoupled from the frontend respectively user-interface.

By now the Tox protocol is an open source project at github.com/TokTok (https://github.com/TokTok/c-toxcore) and a lot of clients for different platforms have been released. All clients are using the same core code of the Tox protocol.

As mentioned in Sect. 4.1, *qTox* [2] is the most popular client for the *Tox* protocol and it is available for different OSs such as Windows, MacOS, and Linux. The first initial commit at *github* was on 24. June 2014 from a user with the canonical name "tux3". In the current version (v 1.16.3) the following features are supported and relevant for a later forensic consideration: Chats, Group Chats, File Transfers, Audio- and Video Calls.

You are able to install *qTox* without registration and without specifying an address, e-mail or your real name. If you start the *qTox* application for the first time, you will be asked to create a profile and provide a username and password or to load an already existing profile. After that, *qTox* generates a unique *Tox ID* and the necessary public and secret keys. This *Tox ID* is your own reachability. You can give this ID to other people to get in contact with them over the tox network.

If you create a new profile, all user information, configuration files and database files are stored at the following locations:

Windows 10:
C:\Users\<username>\AppData\Roaming\tox\.
Linux Ubuntu 18.04.4:
/home/<username>/.config/tox/.

If you set a password while creating an account, all user files and database files are encrypted. The filename is always the *qTox* username. For example, you could have *testuser.db* file for the username "Testuser" and it is the encrypted sqlite3 database. It includes the user contact list and chat history. Another file is the encrypted user configuration file *testuser.ini*, which stores general information about *qTox* user settings.

To read out your own *Tox ID* (cf. 2.2 below), to change your username or user status, you have to open the "My profile" page inside the *qTox* application (Fig. 4.1).

4.2.2 The Tox ID

The unique *Tox ID* is a 76 characters (38 bytes) long string, which is used to identify peers in the *tox* peer to peer network. In order to add a friend, you need to have the friend's *Tox ID*. The first part of the Tox ID is the 32 byte long term public key of the peer. The second part contains a 4 byte *nospam* value and the third part is a 2 byte XOR checksum (Table 4.1).

If a *Tox* user wants to add a friend to his friend-list, he will try to send a friend request to the *Tox ID* of that friend. This friend request contains the asymmetric public key, the *nospam* value and an inviting message. The primary goal of the friend request

Fig. 4.1 "My profile" page inside qTox client on Windows 10

is transmitting the long term public key, because that is what the friend needs to know to create a connection to the sender of the request. The *nospam* value is a number to prevent someone from spamming the peer to peer network with valid friend requests. It makes sure that only those people who have seen the *Tox ID* of a friend are capable of sending a request. The *nospam* value is part of the *Tox ID*.

The 2-byte checksum is calculated by XORing. The first two bytes of the long term public key and *nospam* value are calculated with the next two bytes. This result is calculated with the next following two bytes again until all 36 bytes have been XORed together. Then the result is appended to the end to form the *Tox ID*.

4.3 Related Work

(Instant) Messaging and email applications such as Facebook's chat and Gmail clients, respectively, are widely used communication services that allow individuals to exchange messages over the Internet. Given the nature of the exchanged data, digital artifacts left by such applications may hold highly relevant forensic

Table 4.1 An example of a *Tox ID* structure

32-byte public key	4-byte nospam	2-byte checksum
556DFC51E68F9F41AB9C91CDFBD71AC8E0458CF4507FDCB26AE572B25037E070	5477D564	C2C3

value [11]. Vukadinović [12] described that instant messaging popularity has transitioned from desktop-based-applications (i.e., ICQ, Windows Live Messenger, Yahoo! Messenger) to smartphone-based applications (i.e., WhatsApp, Viber, Kik, WeChat, QQ). There was also a transition from text conversations between individuals to the exchange of media like photos, video, documents, location and audio/video data. The possibilities of instant messaging applications have developed enormously in recent years.

In the past messaging forensics research has been performed and it primarily focused on mobile operating systems like iOS and Android. Gao and Zhang [13] conducted an exploratory study to look for any artifacts left behind by the third most popular IM application worldwide, WeChat on iOS (Statista 2018). They found audio data, conversation databases, user profile information, photo and videos. Yuhang and Tianjie [14] described the forensic analysis of the QQ Tencent messenger, which was the most popularly used messaging application in China at that time. By using digital and memory forensics they were able to extract information like contact lists, chats records, network notepad and display names.

In a series of past works, Dickson identified that artifacts of the client-based AOL messenger (AIM), [15] and Yahoo Messenger [16] could be recovered from the registry and other application files on the hard drive of a Windows XP machine. By using keyword search, Dickson was able to recover artifacts of the conversation history from unstructured datasets such as memory dumps, slack space and swap files in plain text, also if conversation logging was disabled. Thakur [17] analyzed the WhatsApp application on Android to determine what types of user data can be extracted from the external storage and internal memory of the application. He identified that deleted messages in WhatsApp messenger can be extracted from the internal memory of an Android device.

The encryption of user data and profile information is an important and common practice used in messenger applications because diverse type of personal data such as chat messages and user profiles are stored and managed for many different purposes. Kobsa et al. [18] described in their study, that users were concerned about chat message logs that could be abused. There are a few works that deal with the analysis of encrypted databases. Barghouthi and Said [19] analyzed the IM applications Facebook-Messenger, Skype, Gmail, Yahoo and Google Talk to identify the encryption methods used for protecting the conversation through packet sniffer software (like Wireshark) or other network forensic/investigation tools. Choi et al. [20] analyzed the messenger application KakaoTalk and found out, that chat messages can be recovered from the backup files if the user of the application selects a weak password for the backup service.

Researches in the past have shown that different messenger systems leave information, like the IP address of the instant messenger chat session initiator or other personal information, on the system's memory once the application is executed [21]. It was possible in some cases to retrieve the chat history from the volatile memory, when the messenger encrypts the message before sending it over the messenger network [14]. Additional information such as contact lists, group messages, user-names and filenames of sent and received files could also be restored from the volatile memory.

On the other hand, the market of useful forensic software is big. There are some tools available that are able to extract forensic artifacts of well-known messenger like Skype, ICQ and Jabber. An examination of the *qTox* images with the Software Magnet Forensics Internet Evidence Finder (v6.4) was not able to extract relevant artifacts.

The literature research shows that in the past it was essentially possible to find readable artifacts of (instant) messengers on different operating systems with partially different forensic methods. It was easier to extract messenger data, since most of the time no data encryption was used. Time has changed and today most of the well-known messenger applications use profile and database encryption techniques. Experiments in the recent past show that it is mostly only possible to find artifacts by analyzing the memory of a running system, because conversation or other information are readable, before they are stored in encrypted form. Live and memory forensic is becoming more and more important.

4.4   Why qTox Database Forensics?

Recently, many computers have been seized by the police all over the world, but they are not able to get relevant information from the systems regarding *qTox* communication; and as mentioned in Sect. 3, forensic software is not able to restore *qTox* communication and the investigators do not have any clues on how to search for relevant *qTox* data.

The objective of this research is to identify the artifacts stored by *qTox* in the file system and in memory on Windows and Linux. This leads to the following questions:

(a)   What artifacts can forensically be discovered when using *qTox* messenger on Windows and Linux?
(b)   If there are encrypted artifacts of the *qTox* messenger, is there a way for the investigator to decrypt such data?

With regard to research question (a), it is assumed that every software, used on a computer system, leaves its mark on it. With the use of a wide range of forensic analyses, like forensic analysis of a hard drive and forensic analysis of volatile data (memory forensics) it should be possible to reconstruct artifacts from the hard drive and/or the memory of the system. If communication or information of the *qTox* messenger are processed it is always passed to the memory and in certain circum-stances saved on the hard drive. It is expected that data artifacts of the application *qTox* should be found on the hard drive and/or inside the memory. It is always the question how to search for it.

With regard to research question (b) it is possible that there are artifacts that are not readable, because of data encryption. Each profile of the *qTox* messenger uses a

profile database that is encrypted with the profile password. While *qTox* is running
it is necessary to decrypt the profile database, to read the data from the database
or write new data to the database. This is necessary in order to keep the key in the
memory. It is expected that it is possible to find the encryption key inside the memory
dump.

4.5   Methodology

4.5.1   Artifact Definitions

A recoverable artifact is any item of interest recovered from the forensic analysis of
an image or found in a memory dump of a computer system. Regarding the functions
of the *qTox* messenger, there are different types of possible recoverable artifacts as
follows:

- Sent and received text messages
- Deleted text messages
- Contact information about participants of a communication
- *qTox* contacts of the suspect
- Send and received picture, videos and files
- Outgoing and incoming voice call
- Outgoing and incoming video call
- Information about an individual chat conversation
- Information about a group chat conversation
- Information about used keys, hashes and passwords.

The above mentioned types are all related information about the use of *qTox* and
these information are important for further forensic investigations.

4.5.2   Experimental Environments

To search for qTox artifacts on different operating systems, the following environments are set up:

- Linux Host Workstation: Core i7-8700 K CPU @ 3.70 GHz (6 cores), 32 GB
  RAM, Linux Mint 19.3 Tricia with Kernel 5.3.0-51-generic.
- Linux Virtual Environment: Ubuntu-18.04.4, 64-bit, 8 GB RAM, Network: NAT
  Mode. qTox Client v1.17.2 installed with two users namely Alice and Bob.
- Windows Virtual Environment: Windows 10 Enterprise, v1909, Build 18,363.836,
  8 GB RAM, Network: NAT Mode, qTox 64-bit for Windows, v1.17.2 installed
  with one user namely Charlie.

4.5.2.1   qTox Client Setup on Ubuntu 18.04.4

To get the qTox Client (v1.17.2) for Linux clone the qTox source code from https:// github.com/qTox/qTox by using the distributed version-control system Git. Install all related dependencies, listed at https://github.com/qTox/qTox/blob/master/INS TALL.md#ubuntu-git by using the "sudo apt-get command" on the virtual Ubuntu 18.04.4 machine. After that compile the qTox source code by using the cmake command.

qTox Version: 1.17.2-227-g9da1e3bb
Git commit: 0b256c5b83c323a22140ad13bb195d201877a6fa

Start qTox and create a new profile with the following credentials (Table 4.2).

After creating a new profile, there are no contacts in the contact list and no messages or groups are visible. The default profile status is "Toxing on qTox". To have differences here between the clients, change the profile status message to "Alice is toxing on qTox". While creating a new profile, qTox creates some files at the profile location "/home/alice/.config/tox/".

-rw-r-r-1 alice alice 40,960 Mai 25 21:28 **Alice.db**
-rw-r-r-1 alice alice 24 Mai 25 21:28 **Alice.ini**
-rw-r-r-1 alice alice 22 Mai 25 21:28 **Alice.lock**
-rw-r-r-1 alice alice 1084 Mai 25 21:28 **Alice.tox**
drwxrwxr-x 2 alice alice 4096 Mai 25 21:28 **avatars**
-rw-r-r-1 alice alice 6676 Mai 25 21:28 **bootstrapNodes.json**
-rw-r-r-1 alice alice 1531 Mai 25 21:28 **qtox.ini**

When using a password to create a new profile, the files Alice.db (sqlite database) and Alice.tox (profile configuration file) are encrypted by default (Table 4.3 and Fig. 4.2).

After creating the profile, qTox automatically searches for additional peers with whom it can connect. For the first connection, qTox uses fixed bootstrap nodes from a public tox bootstrap nodes list (https://nodes.tox.chat).

The setup of the second Linux Ubuntu machine (Machine 2) for Bob is in the same manner like the first Linux Ubuntu machine for Alice. The profile parameter are different and listed in Table 4.4.

Table 4.2 Credentials for qTox on Linux Ubuntu

Username	Password	Machine
Alice	P4SSAL1C3	Ubuntu 18.04.4 Machine 1
Bob	P4SSBOBX	Ubuntu 18.04.4 Machine 2

Table 4.3 Alice's profile configuration

Alice	Ubuntu 18.04.4 Machine 1
Tox ID	38816B1049BCA73400DC89DB14C98E543B953B69D DFC587211548B45A139587062DE05B42FC3
My name	Alice
My status	Alice is toxing on qTox
Profile Location	/home/alice/.config/tox/
Language	American English
Auto-start	Disabled
Check for updates	Enabled
Default directory to save files	/home/alice/
Auto-accept files	Disabled
Max auto-accept file size	20 MB
Keep chat history	Enabled

Fig. 4.2 Alice's "MyProfile" page at qTox

Table 4.4 Bob's profile configuration

Bob	
	Ubuntu 18.04.4 Machine 2
Tox ID	658ECFA1932BC7F30D00E78F44466572F3A353E82BCC74E30A19AD5A0CF2E254C3616B4C2BE0
My name	Bob
My status	Bob is toxing on qTox
Profile location	/home/Bob/.config/tox/
Language	American English
Auto-start	Disabled
Check for updates	Enabled
Default directory to save files	/home/Bob/
Auto-accept files	Disabled
Max auto-accept file size	20 MB
Keep chat history	Enabled

4.5.2.2   qTox Client Setup on 6.1.5 Windows 10 Ent

The third virtual machine is a Windows 10 Enterprise with the following credentials (Table 4.5).

To install qTox on a Windows 10 operating system download the latest qTox 64-bit executable from https://tox.chat/download.html. The current version of qTox for Windows is v1.17.2. After downloading, execute the file "setup-qtox-x86_64-release.exe" and follow the install instructions (using the default install path: "C:\Program Files\qTox").

Creating the qTox profile in Windows 10 is in the same way as described previously. The profile parameter are different and listed in Table 4.6.

QTox creates the following files at the profile location "C:\Users\Charlie\Roaming\tox" during the profile creation process:

03.06.2020 12:07 < DIR > **avatars**
03.06.2020 12:07 40.960 **Charlie.db**
03.06.2020 12:07 26 **Charlie.ini**
03.06.2020 12:07 64 **Charlie.lock**
03.06.2020 12:13 3.242 **Charlie.tox**
03.06.2020 12:05 369 **qtox.log**

4.5.3   Data Population

To make a forensic analysis and acquire any artifact, a population of data had to take place on the experimental environments. These fiction users *Alice, Bob and Charlie* send different messages to each other including texts and images. There are also group chats with texts and images exchanged from these users.

Following the prepared fictional conversation between the three characters (Table 4.7).

After the text conversation between Alice, Bob and Charlie, three images were exchanged between the parties (Table 4.8).

The third part of communication is a group chat between Alice, Bob and Charlie. To start a group chat in qTox, you have to create a new group and invite friends, who are currently online, from your friendlist to the created group. It is not possible to invite offline friends or external non Tox users to a group chat (Table 4.9).

It is also possible to initiate a voice call inside groups, video calls are not implemented in the current qTox versions, but has already been announced for the next versions. Initiated group sessions are persistent. After closing and restarting the qTox application, initiated groups should be available furthermore.

Table 4.5 Credentials for qTox on Windows 10 enterprise

Username	Password	Machine
Charlie	P4SSCH4RL13	Windows 10 enterprise v1909

Table 4.6 Charlie's profile configuration

Charlie	
	Windows 10 Ent. Machine 3
Tox ID	01501DD783DBC3CDAAA1BDE0818CFEEA81CCF5B3E92F975BB88BCD56E0D8AE05F50A6A0254E39
My name	Charlie
My status	Charlie is toxing on qTox
Profile location	C:/Users/Charlie/AppData/Roaming/tox
Language	American English
Auto-start	Disabled
Check for updates	Enabled
Default directory to save files	C:/Users/Charlie/
Auto-accept files	Disabled
Max auto-accept file size	20 MB
Keep chat history	Enabled

Table 4.7 Fictional conversation between Alice, Bob and Charlie

From	To	Message
Alice	Bob	Hello Bob. How are you?
Bob	Alice	Hello Alice. Here is Bob. I feel very well. And you?
Alice	Bob	Yes Bob. Me too. How is the weather in your city?
Bob	Alice	The sun is shining the weather is sweet :-)
Alice	Charlie	Hello Charlie. Do you come with me to Bob? The sun is shining there and the weather is sweet
Charlie	Alice	Hey Alice. Yes I go with you to Bob. But first I ask Bob if I can come too
Charlie	Bob	Good morning Bob, can I come to you with Alice?
Bob	Charlie	Yes, of course. I am happy to see you later
Charlie	Bob	Perfect. I am also happy to see you Bob
Alice	Bob	Starting audio call. Bob accepted
Bob	Charlie	Starting audio call. Charlie accepted
Bob	Alice	Starting video call. Alice accepted

4.5.4 Acquisition of Data

There are two level of data acquisition: memory and image. For the memory acquisition, memory dumps were carried out for each environment by using *dumpvmcore* command from Virtualbox. The .vdi and/or .vmdk disk image files of each virtual machine were also taken for the image acquisition.

4.5.4.1 Memory Acquisition

Using VirtualBox as a virtual environment has the advantage that some helpful functions are included. A direct memory acquisition is possible with just one virtualbox command. It is also recommended to use these virtualbox functions, because then there is no change of data due to the installation of a forensic tool.

To acquire such data, first a memory acquisition was done in each virtual environment directly after the population of data with the following commands[1] on the host system.

Memory Dump of Machine 1 (Linux Ubuntu)

vboxmanage debugvm "Ubuntu-18.04.4_Machine_1_Alice" dumpvmcore –filename Memory_Ubuntu-18.04.4_Machine_1_Alice.elf

---

[1] Vboxmanage debugvm dumpvmcore: https://www.virtualbox.org/manual/ch12.html#ts_guest-core-format.

Table 4.8 Image exchange between Alice, Bob and Charlie

From	To	Image
Alice	Bob	 Filename: cat.jpg Type: JPEG MD5: 28a174190b533f5f463c7567c14134c3y SHA256: a874d21f13d2f5d61e3138a7b55ad80a55aae6a6152ab56eb88ca86c94a1fbd3
Bob	Charlie	 Filename: dog.jpg Type: JPEG MD5: 77da9f27c4f3722f737a52ee37e2df06 SHA256: 7e6c810ef9c8e3be76c6c2af0295052cd01221285e813abd2bfdf0b1a86719b8
Charlie	Alice	 Filename: mouse.jpg Type: JPEG MD5: 7a210e73e3e6ffe2537e15ace5569b21 SHA256: e4a6e0653bacbab1a6961ed336458a403100a139d232a72c2c43efa889c898a7

Table 4.9 Group chat between Alice, Bob and Charlie

From	To	Message
Alice	Group-chat	Hey Bob and Charlie. I am looking forward to meet you
Bob	Group-chat	Hello Alice and Charlie. Should I cook something?
Charlie	Group-chat	Oh yes. I can bring something to drink
Bob	Group-chat	That would be very awesome Charlie
Alice	Group-chat	Ok guys, see you later. :-)
Alice (sends an image to the group chat)	Group-chat	Filename: swan.jpg Type: JPEG MD5: ed26e2235917c7e60dfb10b43d1ed9a6 SHA256: f4cedc91c863fac9fa2782e754d610f850edee43603939617a06c418303ed26b

Result:

-rwxrwxrwx 1 dm dm 8,2G Jun 4 18:27 Memory_Ubuntu-18.04.4_Machine_1_Alice
.elf

Memory Dump of Machine 2 (Linux Ubuntu)

vboxmanage debugvm "Ubuntu-18.04.4_Machine_2_Bob" dumpvmcore –filename
Memory_Ubuntu-18.04.4_Machine_2_Bob.elf

Result:

-rwxrwxrwx 1 dm dm 8,2G Jun 4 18:27 Memory_Ubuntu-18.04.4_Machine_2_Bob
.elf.

Memory Dump of Machine 3 (Windows 10).

vboxmanage debugvm "UWin10_Ent_Machine_3_Charly" dumpvmcore –filename
Memory_Win10_Ent_Machine_3_Charly.elf

Result:

-rwxrwxrwx 1 dm dm 8,2G Jun 4 18:27 Memory_Win10_Ent_Machine_3_Charly.elf
ELF stands for Executable and Linkable Format and is a common standard file
format for executable files, object code, shared libraries and core dumps.[2]
The ELF files are raw memory dumps and are directly used for further analysis.

4.5.4.2 Image Acquisition

After the population of data and getting the memory dumps via virtualbox functions,
it is necessary to acquire a forensic copy of the hard drives. Because of virtualbox it
is enough to copy the vdi or vmdk disk image files of each virtual machine.

-rwxrwxrwx 1 dm dm 9,1G Jun 4 18:30 Alice_Ubuntu-18.04.4.vdi
-rwxrwxrwx 1 dm dm 9,2G Jun 4 18:30 Bob_20200520_Ubuntu-
18.04.4[…].vmdk
-rwxrwxrwx 1 dm dm 18G Jun 4 18:31 Charlie_Win10_Pro_clean-disk001.vmdk

The vdi and vmdk files can be used in the same way as an image created with
tools like dd or FTK Imager. It acts like a raw copy of the hard drive of the virtual
machines.

---

[2] ELF file format: https://en.wikipedia.org/wiki/Executable_and_Linkable_Format.

4.5.5 Forensic Analysis of Data

When starting with the forensic work, it is helpful to prepare a list with relevant
keywords to get a first impression of relevant matches. Use names, parts of text
messages, filenames and other relevant values and meta data from the data population
to generate a list for the "X-Ways Simultaneous Search Engine".

noindent**Starting list of search terms**

Bob
Charlie
cat.jpg
mouse.jpg
dog.jpg
swan.jpg
10.11.12.46
10.11.12.47
10.11.12.37
< p class = alert >
< p class = msg >
P4SSAL1C3
P4SSBOBX
P4SSCH4RL13
is toxing on qTox

This list of search terms should be dynamic and can be expanded as a part of the
investigation or examination.

Besides, one important part to get forensic artifacts from the *qTox* client is to
analyze the *qTox* sqlite database file for each client profile. By default, the sqlite
database file (*.db) and the *qTox* profile file (*.tox) are encrypted. The database is
encrypted by using SQLCipher. The encryption key is a 256-bit hex value and it can be
derived from the password by using *deriveKey* function in the *qTox rawdatbase.cpp*
at *src/persistence/db/rawdatabase.cpp* (Fig. 4.3).

If the profile password is known, it is possible to extract the relevant profile
files from the evidence image. Put the files in the *qTox* profile directory (e.g.
/home/investigator/.config/tox) of a fresh *qTox* installation and start it. It is possible
to load the relevant profile with the known password and remove the password at the
"My Profile" page.

After removing the password, the sqlite database profile file is no longer encrypted
and could be read with every sqlite3 tool. Figure 4.4 illustrates the Structure of
qTox sqlite database file. Compared to other well know messenger applications like
WhatsApp on mobile devices or Skype, the database schema of *qTox* is very slim.
Not all activities like voice and video calls are stored to the database. Profile status
messages or display name history entries are not saved as well.

```
/**
 * @brief Derives a 256bit key from the password and returns it hex-encoded
 * @param password Password to decrypt database
 * @param salt Salt to improve password strength, must be TOX_PASS_SALT_LENGTH bytes
 * @return String representation of key
 */
 QString RawDatabase::deriveKey(const QString& password, const QByteArray& salt)
 {
if (password.isEmpty()) {
  return {};
}

if (salt.length() != TOX_PASS_SALT_LENGTH) {
  qWarning() << "Salt length doesn't match toxencryptsave expections";
  return {};
}

const QByteArray passData = password.toUtf8();

static_assert(TOX_PASS_KEY_LENGTH >= 32, "toxcore must provide 256bit or longer keys");
const std::unique_ptr<Tox_Pass_Key, PassKeyDeleter> key(tox_pass_key_derive_with_salt(
  reinterpret_cast<const uint8_t*>(passData.data()),
  static_cast<std::size_t>(passData.size()),
  reinterpret_cast<const uint8_t*>(salt.constData()), nullptr));
return QByteArray(reinterpret_cast<char*>(key.get()) + 32, 32).toHex();
  }
```

Fig. 4.3 *deriveKey* function in *rawdatbase.cpp*

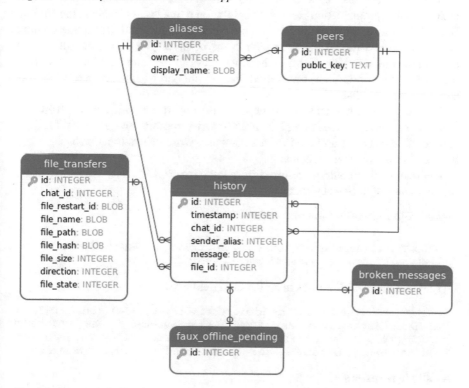

Fig. 4.4 Structure of *qTox* database file

Only direct communication between peers are saved to the database. Group conversations or information about group members like display name or the public key are not saved to the database either. Each default *qTox* profile has six tables. The tables "*faux_offline_pending*" and "*broken_messages*" were not considered, because they are not important to restore relevant communication.

The "*peers*" table contains the *Tox IDs* of the profile owner and the communication partners (peers) in the column "*public_key*". The public key is the first 32 byte of the *Tox ID*. The ID field of the table is a unique (auto increment) value and serves as a relationship ID to other tables.

The "*aliases*" table contains the display names of the profile owner and the communication partners in the column "*display_name*" and the related peer ID from the "*peers*" table in the column "*owner*".

The "*history*" table contains all messages sent and received with information about the message content in the column "*message*" and the communication partner ID in relation to the "*peers*" table in the column "*chat_id*". It also contains the used sender alias ID in relation to the "*alias*" table, the timestamp of the message in the unix timestamp format in milliseconds and the "*file_id*" in relation to the "*file_transfers*" table, if a file was sent.

The "*file_transfers*" table contains all files that have been sent or received. It contains the communication partner ID in relation to the "peers" table in the column "*chat_id*". It also contains the original file name of transferred files in the column "*file_name*", the original local file path in the column "*file_path*", the file size in bytes in the column "*file_size*" and the direction of the transfer (sent or received) in the column "direction". The transferred files are not part of the database, rather it is saved on the hard disk.

In order to communicate with the database and to get information from the database the query language SQL is used to write appropriate queries. With SQL it is possible to get information from a database, write values to a database, delete data or alter the database schema.

Communication could be restored from the qTox SQLite database file by using the following SQL Select statements:

Get all communication partners/peers

```
SELECT aliases.display_name, peers.public_key FROM aliases
INNER JOIN peers ON aliases.owner=peers.id;
```

Get all messages with names in a chronological order

```
SELECT     datetime(substr(history.timestamp,0,11),'unixepoch'),
aliases.display_name as 'FROM', history.chat_id as 'TO CHAT
ID',history.message FROM history INNER JOIN aliases ON
aliases.owner=history.sender_alias ORDER by history.timestamp;
```

Get all file transfers

```
SELECT chat_id,file_name,file_path,file_size,direction,
file_state FROM file_transfers;
```

Get all messages with names and file transfers in a chronological order

```
SELECT    datetime(substr(history.timestamp,0,11),'unixepoch'),
aliases.display_name as 'FROM', history.chat_id as 'TO CHAT ID',
history.message,file_transfers.file_name,file_transfers.
file_path,file_transfers.file_size,file_transfers.file_state
FROM history INNER JOIN aliases ON aliases.owner=history.
sender_alias LEFT JOIN file_transfers ON history.file_id=file
_transfers.id ORDER by history.timestamp;
```

Explanation "chat_id":
The column "chat_id" describes the unique ID of a communication channel between two parties. If Alice establishes a new conversation by sending an invitation to another Tox ID, the "chat_id" increases by one. All messages and files are sent to a communication channel.

The timestamp of the *history* table is a unix timestamp in millisecond format, saved as an integer value inside the database. To make it readable and for a further analysis it is helpful to convert the unix timestamp to a standardized time format. This is possible inside the SQL statement.

4.6   Findings and Discussions

This section presents the artifacts found on each virtual machine after conducting the forensic analysis with Autopsy and a manual analysis of the *qTox* database file. The evidence image file, the memory dump and the database file are considered for each environment.

4.6.1   Recovered Artifacts Found in the Image Files

qTox debug log files
qTox *log* files are found in both Linux and Windows. The file structure in the different profile directories is almost identical.

The file *"qtox.log"* is not encrypted and available on Linux and Windows in the respective profile directory and contains debug information regarding the execution of *qTox*. Activities found in the debug log are documented with the exact timestamp and can, under certain circumstances, be used to enrich user activities and *qTox* conversations. From the log files, it is possible to find information about the following user activities during the use of *qTox* (Tables 4.10, 4.11, 4.12 and 4.13):

- Artifacts about friend requests
- Artifacts about sending and receiving files
- Artifacts about audio and video calls
- Artifacts about group conversation and attendees of a group.

Table 4.10 Artifacts found in debug log file/receiving a file request

File: qtox.log	Info: qTox debug log file/Charlie/Windows
Action	Bob sent the file "dog.jpg" to Charlie @ 2020-06-04 18:10:51 (UTC: 2020-06-04 16:10:51)
Finding	[16:10:51.930 UTC] core/corefile.cpp:357: Debug: "filename already clean" [16:10:51.930 UTC] core/corefile.cpp:359: Debug: "Received file request 1:65,536 kind 0"
Explanation	It is recognizable, that Charlie received a file request at the same time Bob sent a file to Charlie

Table 4.11 Artifacts found in debug log file/sending file request

File: qtox.log	Info: qTox debug log file/Charlie/Windows
Action	Charlie sent the file "mouse.jpg" to Bob @ 2020-06-04 18:11:30 (UTC: 2020-06-04 16:11:30)
Finding	[16:11:30.220 UTC] core/corefile.cpp:158: Debug: "sendFile: Created file sender 0 with friend 1"
Explanation	It is recognizable, that Charlie sent a file request to Bob

Table 4.12 Artifacts found in debug log file/making audio call

File: qtox.log	Info: qTox debug log file/Charlie/Windows
Action	Bob initiate an audio call with Bob @ 2020-06-04 18:21:41 (UTC: 2020-06-04 16:21:41)
Finding	[16:21:41.728 UTC] audio/backend/openal.cpp:441: Debug: Opening audio output "Lautsprecher (High Definition Audio Device)" [16:21:41.728 UTC] audio/backend/openal.cpp:453: Debug: Opened audio output "Lautsprecher (high definition audio device)" [16:21:41.773 UTC] audio/backend/openal.cpp:278: Debug: Audio source 1 created. Sources active: 1 [16:21:41.773 UTC] core/coreav.cpp:732: Debug: "Received call invite from 1"
Explanation	It is recognizable, that Bob initiate an audio call with Charlie

Table 4.13 Artifacts found in debug log file/group conversation. Action describes an action in the context of data population

File: qtox.log	Info: qTox debug log file/Charlie/Windows
Action	Charlie joins the group "friends" @ 2020-06-04 18:15:55 (UTC: 2020-06-04 16:15:55)
Finding	[16:15:55.328 UTC] core/core.cpp:600: Debug: "Group 0 peerlist changed" [16:15:55.328 UTC] core/core.cpp:609: Debug: "Group 0, Peer 0, name changed to Alice" [16:15:55.328 UTC] core/core.cpp:600: Debug: "Group 0 peerlist changed" [16:15:55.328 UTC] core/core.cpp:609: Debug: "Group 0, Peer 1, name changed to Bob" [16:15:55.328 UTC] core/core.cpp:600: Debug: "Group 0 peerlist changed"
Explanation	The debug log file shows information about the group members and the peer IDs at the time Charlie joins the group

The created log file contains relevant information about user activities. None of the information like audio and video calls or group activities are included in the profile database. The entries are logged with the exact timestamp in UTC (coordinated universal time).

Receiving a file request:
Sending a file request:
Making an audio call:
Group conversation:

Conversations

Artifacts or communication extracts about accomplished conversations between Alice, Bob and Charlie were not found on the image files. Primarily it is because of the *qTox* database encryption. Conversations between peers are processed in the system's memory and then written to the database. The database is encrypted again after access. There is no situation, when *qTox* writes plain text from conversations to the hard disk.

The transferred files were found in the respective user directory of the system:

- Linux machine 1 (Alice): "mouse.jpg" was found at **/home/alice/Pictures /mouse.jpg**
  The hash values are identical to hash values of the source files.
  MD5: 7a210e73e3e6ffe2537e15ace5569b21
  SHA256: e4a6e0653bacbab1a6961ed336458a403100a139d232a72c2c43 efa889c898a7
- Linux machine 2 (Bob): "cat.jpg" was found at **/home/dm/Pictures/cat.jpg**
  The hash values are identical to hash values of the source files.
  MD5: 28a174190b533f5f463c7567c14134c3y
  SHA256: a874d21f13d2f5d61e3138a7b55ad80a55aae6a6152ab56eb88 ca86c94a1fbd3
- Windows machine 3 (Charlie): "dog.jpg" was found at **\Users\spam\Pictures \dog.jpg**
  The hash values are identical to hash values of the source files.
  MD5: 77da9f27c4f3722f737a52ee37e2df06
  SHA256: 7e6c810ef9c8e3be76c6c2af0295052cd01221285e813abd2bfdf 0b1a86719b8

Files that are transferred during the conversation are not saved in the database. In the database table *"file_transfers"*, there is only a reference to the physical location of the transferred file located. The transferred files were not changed or compressed. The hash values of the received files were identical to the hash values of the source files.

4.6.2   Recovered Artifacts Found in the Memory Dump

Extraction of text conversation

The search for relevant text conversation relating to the data population [4.2] gives the below mentioned results. It is possible to reduce the amount of false positive results by using the following search terms to extract artifacts of qTox text conversation from the memory dumps:

    < p class = alert >
    < p class = msg >

QTox conversation inside the application is formatted via html tags. The a/m html tags are typical for qTox conversations and helpful, when searching in unstructured raw data.

Table 4.14 shows the extraction results of text conversation from the memory dump for the different virtual machines.

VM 1 = Linux Ubuntu Machine 1 (Alice)
VM 2 = Linux Ubuntu Machine 2 (Bob)
VM 3 = Windows 10 Machine 3 (Charlie)
A = Alice
B = Bob
C = Charlie
G = Group conversation
F = From
T = To

It was possible to extract artifacts of all text messages of the data population (cf. 5.3) from the memory dumps of the different relevant virtual machines.

qTox conversations are processed inside the memory of the *qTox* application and, in case of a peer to peer conversation, saved to the database. Group conversations are not saved to the database, but also processed inside the memory. All group conversations could be read out completely from the memory dump.

Each message (peer to peer and group conversations) could be found multiple times at different offsets in the memory dump. This happens because each message is processed multiple times inside the application and written to different variables as it is received, processed, displayed and stored.

Extraction of qTox encryption keys

As mentioned above (cf. 5.5), the sqlite database of the qTox user profile encrypted by default. *SQLCipher* with some PRAGMA statements is used to ensure the encryption and decryption of the database content.

When the memory dump was created as the qTox client was started, it is possible to get the hex value of the encryption key. During tests it was possible to get the "magic number (hex value)" to find the relevant encryption key as a hex value inside

Table 4.14 Artifacts extraction of text messages

F	T	From Data Population	VM 1	VM 2	VM 3
A	B	Hello Bob. How are you?			
B	A	Hello Alice. Here is Bob. I feel very well. And you?			
A	B	Yes Bob. Me too. How is the weather in your city?			
B	A	The sun is shining the weather is sweet ;-)			
A	C	Hello Charlie. Do you come with me to Bob? The sun is shining there and the weather is sweet.			
C	A	Hey Alice. Yes i go with you to Bob. But first i ask Bob if i can come too.			
C	B	Good Morning Bob, can i come to you with Alice?			
B	C	Yes, of course. I am happy to see you later.			
C	B	Perfect. I am also happy to see you Bob.			
F	**T**	**Group conversation**			
A	G	Hey Bob and Charlie. I am looking forward to meet you.			
B	G	Hello Alice and Charlie. Should i cook something?			
C	G	Oh yes .. I can bring something to drink.			
B	G	That would be very awesome Charlie.			
A	G	Ok guys, See you later. :-)			

unstructured raw data. QTox encrypts the database with the hex value of the profile password. Therefore, encrypted qTox databases can also be decrypted again by using the hex key.

Artifacts of the encryption key can be resolved from unstructured raw data by searching for the following hex combination (Magic Key) inside the memory dump:

6800 0000 0000 0000 7827

The hex value of the database encryption key could be determined for all virtual machines. Figures 4.5 illustrates the extracted encryption key of Alice (96 bytes right after the Magic Key), which is *2712f58e7e09596012e59b7048bcbb4d2a99ce3e8e79b93a4f05537fef3960d05e764 ca1d2e2aafe162ef2f471a4783d* (its hex value is "32 37 31 32 66 35 38 65 37 65 30 39 35 39 36 30 31 32 65 35 39 62 37 30 34 38 62 63 62 62 34 64 32 61 39 39 63 65 33 65 38 65 37 39 62 39 33 61 34 66 30 35 35 33 37 66 65 66 33 39 36 30 64 30 35 65 37 36 34 63 61 31 64 32 65 32 61 61 66 65 31 36 32 65 66 32 66 34 37 31 61 34 37 38 33 64"); and Fig. 4.6 illustrates the extracted key of Charlie (96 bytes right after the Magic Key), which is *86784bd9554c0d3a5636793da578dd93f5df5ba4831bf497be7c448d8c95c207f3963 ebfe072d3bc45a2a24dfa3927a5* (its hex value is "38 36 37 38 34 62 64 39 35 35 34 63 30 64 33 61 35 36 33 36 37 39 33 64 61 35 37 38 64 64 39 33 66 35 64 66 35 62 61 34 38 33 31 62 66 34 39 37 62 65 37 63 34 34 38 64 38 63 39 35 63 32 30 37 66 33 39 36 33 65 62 66 65 30 37 32 64 33 62 63 34 35 61 32 61 32 34 64 66 61 33 33 39 32 37 61 35").

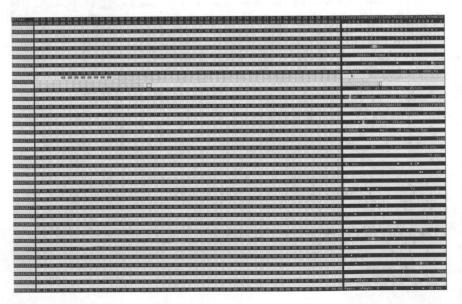

Fig. 4.5 Found location of the encryption key of Alice (the header hex is marked in yellow)

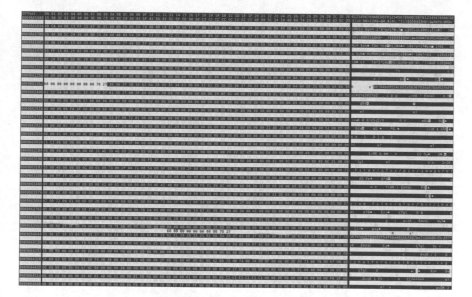

Fig. 4.6 Found location of the encryption key of Charlie (the header hex is marked in yellow)

The encryption keys are used to encrypt and decrypt the *qTox* profile databases. By using SQLCipher + the encryption keys it is possible to decrypt the sqlite database from a forensic image.

Extraction of qTox status text

Artifacts of the qTox status text of all friends in the buddy list can be restored from the different memory dumps. The qTox status message is not stored to the profile database. The relevant strings were found several times in the memory dump. The status message of the profile owner and the peers of the qTox user could be extracted.

4.6.3 Recovered Artifacts Found in the Database Files

For each profile created, a sqlite database file is created in the respective profile directory. The file is named with the profile username and the prefix ".db". If the file is encrypted and the password available, it is possible to create an unencrypted version of the database file. To decrypt the database without password, an option is to get the encryption key from a memory dump and decrypt the database file with "SQLCipher".

Table 4.15 shows an example of messages extracted from the databases and Table 4.16 shows the artifacts from the file transfer extracted from the database.

Table 4.15 Get messages from database file

#	Timestamp	From	Chat_id	Message
1	2020-06-04 15:53:33	Alice	0	/me offers friendship, "Alice here! Tox me maybe?"
2	2020-06-04 15:56:16	Alice	2	/me offers friendship, "Alice here! Tox me maybe?"
3	2020-06-04 15:57:25	Alice	3	/me offers friendship, "Alice here! Tox me maybe?"
4	2020-06-04 16:02:25	Alice	2	Hello Bob. How are you?
5	2020-06-04 16:03:15	Bob	2	Hello Alice. Here is Bob. I feel very well. And you?
6	2020-06-04 16:03:47	Alice	2	Yes Bob. Me too. How is the weather in your city?
7	2020-06-04 16:05:01	Bob	2	The sun is shining the weather is sweet ;-)
8	2020-06-04 16:06:44	Alice	3	Hello Charlie. Do you come with me to Bob? The sun is shining there and the weather is sweet
9	2020-06-04 16:07:31	Charlie	3	Hey Alice. Yes I go with you to Bob. But first I ask Bob if I can come too
10	2020-06-04 16:10:03	Alice	2	
11	2020-06-04 16:11:58	Charlie	3	

Table 4.16 Get file transfers from database file

#	Chat_id	File_name	File_path	File_size	Direction	State
1	2	cat.jpg	/run/user/1001/gvfs/afp-volume:host = nas.local,user = dm,volume = Data/.../cat.jpg	41,629	0	5
2	3	mouse.jpg	/home/alice/Pictures/mouse.jpg	52,023	1	5

4.6.4 Discussion

The experiment shows, that it is possible to recover several artifacts from the test environment by using different forensic techniques. The first was to analyze the virtual machine images, the second was to analyze the memory dump of each machine and the third was to investigate the qTox profile database file. Different artifacts were found with each technique. With all the techniques together, almost all of the information were restored.

All recovered messages matched the data population. Specifically, it was possible to recover 9 of 9 text messages between Alice, Bob and Charlie from the experiments. Status messages like "invitation" were also restored as well as 5 of 5 group messages. Audio calls between the communication partners were traced and the profile status

Table 4.17 Results of data extraction

Input	Amount	#Artifact found
Text messages	9	9
Text messages to group	5	5
Ansfered images	4	3
Audio calls	3	3
Communication partners	3	3
Profile status text	3	3
Database encryption keys	3	3

text was resolved. With the discovery of the encryption keys, it was possible to decrypt the *qtox* databases (Table 4.17).

The experiments could prove that with the help of forensic techniques and methods, it was possible to restore artifacts from the images and memory dump files. In the experiment environment, it was able to restore all peer-to-peer messages from the sqlite database of the *qTox* profile and also from the memory dump. Group conversations are not stored to the database file, but could be restored via the memory dump.

Concerning the examination of the memory dump, it depends on the timing between the conversation and the creation of the memory dump file. If the time span is too long, it is possible that memory area has been freed and used by other applications. Existing artifacts could then be overwritten. In the experiments the time span was short enough to proof, that artifacts could be extracted from the memory dump.

In order to extract conversations from the dump file, it was helpful to find and define unique tags to mark relevant *qTox* messages in unstructured raw data. This is also helpful for further investigations in a relevant *qTox* case.

Besides the conversations, information about peers and file transfers were restored from the sqlite database. Transferred files are not stored in the database themselves. An examination of the hard disk image is therefore always necessary here in order to save the associated file. In the experiments all transferred files could be restored from the hard disk image. The transferred files were not changed during the transfer process over the peer-to-peer network because the hash values of the files did not change.

Debug logging of the *qTox* application is activated by default. Because of this, it was possible to extract and analyze the generated log file with relevant information on user activities. In combination with the other forensically obtained artifacts, it is possible to create a detailed picture of user activity.

Regarding the encrypted artifacts of the *qTox* messenger, it was possible as a part of the experiment to locate the used encryption key inside the memory dump. In this context it was helpful to find a "magic number (hex value)" to extract the key for future investigations under different operating systems. With this key and without knowing the real password it was possible to decrypt the sqlite profile database.

4.7 Summary

As already mentioned, it was able to find and restore forensic artifacts under the given circumstances. Basically the results are useful for investigators, because it shows a way to examine a system with a running *qTox* client and it shows first clues where an investigator can start to search and which search terms are reasonable (html tags and hex values) to look inside the digital image and memory dump.

From a technical point of view, there are some limitations. Everything took place in an experimental setting. In the area of memory forensic in particular, there can be many factors what would have a negative impact on the result.

For this analysis the main operating systems Windows 10 and Ubuntu Linux were used. This should cover most of the cases. It was not possible to use a MacOS operating system, because of availability reasons. But it should not differ significantly from the Linux results.

qTox calls itself a secure end-to-end encrypted messenger. This paper has proven that the *qTox* messenger leaves artifacts on a system that help investigators with their work. It makes sense to look deeper and search for possibilities to extract relevant information. *qTox* is being further developed. It makes sense to keep track of changes to the source code and to document the effects on forensic work.

The peer-to-peer network properties of the *tox* protocol offer further approaches for identifying and locate suspects. That could not be tested in this research and could be the subject of future work.

References

1. Schipper, G. C., Seelt, R., & Le-Khac, N.-A. (2021). Forensic analysis of matrix protocol and Riot.im application. *Forensic Science International: Digital Investigation, 36*(Supplement), 301118. https://doi.org/10.1016/j.fsidi.2021.301118.
2. https://qtox.github.io/
3. Cents, R., & Le-Khac, N.-A. (2020). Towards a new approach to identify WhatsApp Messages. In *19th IEEE International Conference on Trust, Security and Privacy in Computing and Communications (IEEE TrustCom-20)* (CORE Rank A).
4. Terrelonge, L. (2017). *Cybercrim economy: An analysis of criminal communications strategies.* [Online]. https://www.forensicfocus.com/stable/wp-content/uploads/2017/05/flashpoint_cybe rcrime_economy.pdf. Accessed March 2022.
5. Wijnberg, D., & Le-Khac, N.-A. (2021). Identifying interception possibilities for What-sApp communication. *Forensic Science International: Digital Investigation, 38*(Supplement), 301132. https://doi.org/10.1016/j.fsidi.2021.301132
6. Wendzel, S., Mazurczyk, W., Caviglione, L. (2019). Advanced information hiding techniques for modern botnets. In G. Kambourakis, et al. (Eds.) *Botnets: Architectures, countermeasures, and challenges.* CRC Press. https://doi.org/10.1201/9780429329913-4
7. https://pnpacg.ph/main/press-releases/2-uncategorised/349-acg-cyber-security-bulletin-nr-198-beware-of-ragnar-locker-ransomware
8. https://www.magnetforensics.com/products/magnet-ief/
9. https://www.magnetforensics.com/products/magnet-axiom/
10. http://www.x-ways.net/forensics/

11. Barradas, D., et al. (2017). Forensic analysis of communication records of web-based messaging applications from physical memory. In *Proceedings of the 14th International Joint Conference on e-Business and Telecommunications* (Vol. 4, pp. 43–54). SECRYPT.
12. Vukadinovic, N. V. (2019). *WhatsApp forensics: Locating artifcats in web and desktop clients.* M.Sc. Thesis, Purdue University.
13. Gao, F., & Zhang, Y. (2013). Analysis of WeChat on IPhone. In *Proceedings of the 2nd International Symposium on Computer, Communication, Control and Automation.* https://doi.org/10.2991/3ca-13.2013.69
14. Yuhang, G., & Tianjie, C. (2010). Memory forensics for QQ from a live system. *Journal of Computers, 5*(4). https://doi.org/10.4304/jcp.5.4.541-548
15. Dickson, M. (2006). An examination into AOL Instant messenger 5.5 contact identification. *Digital Investigation, 3*(4), 227–237.
16. Dickson, M. (2006). An examination into Yahoo messenger 7.0 contact identification. *Digital Investigation, 3*(3), 159–165.
17. Thakur, N. S. (2013). *Forensic analysis of WhatsApp on Android smartphones.* University of New Orleans Theses and Dissertations, 1706.
18. Kobsa, A., Patil, S., & Meyer, B. (2022). *Privacy in Instant messaging: An impression management model.* [Online]. http://citeseerx.ist.psu.edu/viewdoc/versions? 10.1.1.158.7888. Accessed March 2022.
19. Barghuthi, N. B. A., & Said, H. (2013). Social networks IM forensics: Encryption analysis. *Journal of Communications, 8*(11), 708–715.
20. Choi, J., et al. (2017). Forensic analysis of the backup database file in KakaoTalk messenger. In *2017 IEEE International Conference on Big Data and Smart Computing (BigComp)* (pp. 156–161).
21. Carvey, H. (2004). Instant messaging investigations on a live Windows XP system. *Digital Investigation, 1*(4), 256–260.

Chapter 5
PyBit Forensic Investigation

Benno Krause, Kim-Kwang Raymond Choo, and Nhien-An Le-Khac

5.1 Introduction

The daily work of a cyber-investigator contains data forensic related cases as well as network-related cases. In the last 20 years [1], the internet, and thus the so-called "darknet" comes more into focus. Criminals and suspects are using internet techniques to support their criminal activities. Often their criminal activities involve internet techniques themselves.

While during early years of the worldwide internet, communication was mainly unencrypted, nearly every communication protocol has its encrypted pendent. The usage of unencrypted email messaging has been redeemed by encrypted email transport. Messengers like IRC, ICQ or MSN have been replaced by encrypted messengers like Skype or WhatsApp [2, 3].

Internet users primarily are using existing communication platforms. The fewest users or user collectives are developing their own communication methods. That also applies to criminals and suspects. They use what the market has to offer to perform their criminal activities like digital blackmailing (ransomware [4]) or botnets, where secure communication played an essential role in the behavior of the suspects [5].

There are multiple approaches, law enforcement can investigate if encrypted messaging is used by criminals. Often the first approach is to oblige the responsible operator. Another approach is encryption. If the encryption or its implementation is poor, it can be tried to decrypt the message content. The next one is gathering and analyzing existent metadata. Often identification of a message sender by exploring connected IP addresses is expedient.

Knowing the circumstances, the developers of messaging protocols and software try to prevent these approaches.

A commonly used messenger of the cybercriminals is the PyBitmessage Messenger, which is using the Bitmessage protocol [6–9]. The message-protocol is an encrypted open-source peer-to-peer network protocol, which copes without central infrastructure. PyBitmessage is the most used client for the BitMessage protocol. It is Python-based and stores data to a SQLite database and several JSON Datafiles [10].

The Bitmessage protocol, the PyBitmessage Messenger is based on, has been developed with the aspiration of preventing all of the above-mentioned investigations approaches. Also, cybercriminals have become aware of this messaging protocol. For example, the malware Chimera had used the Bitmessage peer-to-peer messaging application to communicate between the victim's computer and the malware developer's command and control server [11]. This creates a decryption service that is **incredibly** portable, secure, and difficult, if not impossible, to take down as all the peers in the network are helping to distribute the keys. Hence this chapter describes how to gather pieces of information about messages, sender, and recipients of the Bitmessage network, how to analyze PyBitmessage database files with the specific stipulation to support investigations.

5.2 Basic Features

5.2.1 Bitmessage Concept

BitMessage is a system, that allows users to securely send and receive messages using a trustless, decentralized peer-to-peer protocol. The protocol and the encryption are inspired by the Bitcoin protocol.

All messages including content and metadata like recipient and sender, are encrypted. In the Bitmessage protocol, the encrypted message does not reveal any information about the sender or the receiver. Therefore it has to be delivered to all participants of the BitMessage network.

Like common Email encryption techniques like PGP/GPG, a message is encrypted with the public key of the receiver. The receiver is the only one able to decrypt the message, as only he knows the corresponding private key. Every receiver tries to decrypt the message. If he succeeds, he knows that he is the proper receiver.

For the identification of a BitMessage participant, a Bitcoin-like address is used. This address is the hash of participants' public key and a checksum encoded with base58, prepended by the characters "BM-"

BM-2cUTQn8kRiLSJseQgNqtbcMpWSAbboVNRD

5.2.2 Data Encryption

5.2.2.1 Symmetric-Key Cryptography

Message objects are encrypted by using *Symmetric-Key Cryptography*. To encrypt a plain text, or decrypt a ciphertext, symmetric-key algorithms only need one single identical key. The symmetry relies on the reversible calculation of one single secret. This secret-key needs to remain hidden and is never communicated in plaintext.

The symmetric-key cryptography in BitMessage is implemented by using AES-256-CBC. The Advanced Encryption Standard (AES) is an international encryption standard initiated by the National Institute of Standards and Technology for symmetric-key encryption. The AES algorithm is a symmetric block cipher that can encrypt (encipher) and decrypt (decipher) information. Encryption converts data to an unintelligible form called ciphertext; decrypting the ciphertext converts the data back into its original form, called plaintext. The AES algorithm is capable of using cryptographic keys of 128, 192, and 256 bits to encrypt and decrypt data in blocks of 128 bits [12].

Bitmessage uses AES-256-CBC with a key length of 256 bit and a block length of 128 bit.

5.2.2.2 Asymmetric-Key Cryptography

The sender and receiver have to agree on a key before the communication starts. These agreements must be secure. The key must be established between the sender and recipient using a secure channel.

In order to overcome this challenge, BitMessage uses Diffie-Hellmann Key-Exchange. Diffie, Hellmann, and Merkle had a proposal based on the following idea. The key used by the person who encrypts the message doesn't need to be secret. The crucial part is that the receiver can only decrypt using a secret key [13].

Bitmessage uses the Elliptic Curve Integrated Encryption Scheme (ECIES) to encrypt the payload of the Message and Broadcast objects. The scheme uses Elliptic Curve Diffie-Hellman (ECDH) to generate a shared secret used to generate the encryption parameters for the Advanced Encryption Standard with 256-bit key and Cipher-Block Chaining (AES-256-CBC).

The given code examples are taken from the BitMessage reference client PyBitMessage.

5.2.2.3 Public and Private Key Generation in BitMessage

BitMessage uses the Elliptic Curve Digital Signature Algorithm (ECDSA) based on elliptic curve cryptography. The particular elliptic curve is known as secp256k1 (Koblitz Curve), which is the curve $y^2 = x^3 + 7$.

The BitMessage private key is a random value, created by the OpenSSL rand function with a size of 32 bit.[1] The determining of the public key is done by an elliptic curve point multiplication using the secp256k1 curve.[2] The generation of the

---

[1] https://github.com/Bitmessage/PyBitmessage/blob/v0.6/src/class_addressGenerator.py#L138.

[2] https://github.com/Bitmessage/PyBitmessage/blob/v0.6/src/highlevelcrypto.py#L106.

public key has to be repeated, until the first bytes of *RipeMD160(SHA512(private key + public key))* begins with \ × 00 or \ × 00\ × 00.

Key Storing Format

In PyBitMessage, the keys are stored in the Bitcoin "Wallet import format". For this, the key has to be converted:

1. Adding a 0 × 80 byte in front of the key.
2. Creating a checksum of the key by

   (a) performing a SHA-256 hash on the key
   (b) and an additional SHA-256 hash on the hash result
   (c) and taking the first 4 bytes.

3. Adding the 4 bytes checksum at the end of the key.
4. Converting the key into base58.

Example of converting a given private key into the "Wallet import format".

1. The following 32bit private key is given:
   5f:14:bd:dc:df:05:27:28:0f:b2:e9:78:19:a2:4e:87:f6:b5:c7:d6:23:f3:73:2d: d5: a2:a7:f0:1c:1c:0f:fa.
2. Adding a 0 × 80 byte in front of the key.
   80:5f:14:bd:dc:df:05:27:28:0f:b2:e9:78:19:a2:4e:87:f6:b5:c7:d6:23:f3:73:2d: d5:a2:a7:f0:1c:1c:0f:fa.
3. Creating a checksum of the key

   (a) SHA-256 hash of the key
       b8:e1:3e:de:de:64:f7:ba:d5:4a:de:33:23:cc:14:07:d1:a3:d7:67:bd:62:7d: 9a:73:8f:eb:80:85:13:16:48.
   (b) SHA-256 hash of the above SHA-256 hash
       57:29:2a:91:5b:f8:85:af:94:85:c5:3e:ff:89:a3:4c:6b:a8:34:35:af:b8:13: 64:1c:cb:f3:2f:d5:f6:56:14.
   (c) Taking the first 4 Bytes
       57:29:2a:91.

4. Adding the 4 bytes checksum at the end of the key.
   80:5f:14:bd:dc:df:05:27:28:0f:b2:e9:78:19:a2:4e:87:f6:b5:c7:d6:23:f3:73: 2d:d5:a2:a7:f0:1c:1c:0f:fa:57:29:2a:91.
5. Converting the key into base58
   5JYAJGmG58M1v8msZqTop5RxCeYdNattBg8gv3MKKMwzW1MyBZi.

Visualization of the Public Key

The BitMessage public key has a size of 20 byte. It's part of the BitMessage identifier (Bitmessage address).

The following steps are mandatory to convert a public key into its visual representation:

1. Removing the leading zero bytes
2. Merging of [Adressversion] + [Stream number] + [Key]
3. Creating a checksum of the result of step 2 by

   a. performing a SHA-512 hash on the key
   b. and an additional SHA-512 hash on the hash result
   c. and taking the first 4 bytes.

4. Merging the result of step 2 and the checksum of step 3
5. Converrting into base58
6. Adding "BM-"

Example of creating the visual representation of a public key.
BM-2cVpCsorUSNZkK8ENC89C2XCHEneTZxG6J can be created by the following steps:

1. Given is the following public key:
   00:96:bf:d6:ad:6f:67:42:55:9d:7f:e3:a0:d2:11:e6:bb:cd:78:31.
2. Removing the leading zero byte(s)
   96:bf:d6:ad:6f:67:42:55:9d:7f:e3:a0:d2:11:e6:bb:cd:78:31.
3. Merging of [Adressversion] + [Stream number] + [Key]
   [04] + [00] + [96:bf:d6:ad:6f:67:42:55:9d:7f:e3:a0:d2:11:e6:bb:cd:78:31].
       04:00:96:bf:d6:ad:6f:67:42:55:9d:7f:e3:a0:d2:11:e6:bb:cd:7:31.
4. Creating a checksum of the result of Step 2 by

   a. performing a SHA-512 hash on the key
      68:79:f0:70:d9:e1:64:a9:dc:7f:29:09:5a:e6:69:b0:f2:78:11:65:a8:
      e6:ac:66:73:58:1f:e2:fb:30:93:50:2e:fe:b6:48:3c:49:48:38:e5:73:3c:16:51:
      90: ca:42:ca:13:15:a6:0a:c5:18:1c:e2:07:5b:4c:3a:e5:be:cb.
   b. and an additional SHA-512 hash on the hash result
      a5:29:11:e7:d3:13:02:a0:5c:8a:a8:cf:fa:29:54:2b:64:a4:c2:53:2d: 54:3e:6b:
      2e:8d:7a:47:1d:02:5d:ba:8c:0f:43:c1:9a:81:cf:25:fd:96:b2:c9:27:   33:38:
      7a:0d:3a:e4:e1:07:f6:a4:c8:b3:67:96:4f:57:2d:81:b7.
   c. and taking the first 4 bytes.
      a5:29:11:e7.

5. Merging the result of step 2 and the checksum of step 3
   04:00:96:bf:d6:ad:6f:67:42:55:9d:7f:e3:a0:d2:11:e6:bb:cd:7:31:a5:29:11:e7.
6. Converting into base58
   2cVpCsorUSNZkK8ENC89C2XCHEneTZxG6J.
7. Adding "BM-"
   BM-2cVpCsorUSNZkK8ENC89C2XCHEneTZxG6J.

5.2.3 BitMessage Networking Concept

The concept of messaging is highly orientated on BitCoin and PGP. The message and it's metadata has to be encrypted by the public key of the recipient and signed by the private signing key of the sender. The messages are propagated though the whole network, meaning that every peer in the network gets every message. To send a message to another person (receiver), the message is passed from peer to peer until it reaches the recipient. Each peer repeatedly download messages from a neighbor peer. Stored objects are again downloaded from other peers and so on, until the receiver downloads the message from another peer he is connected to. Only the receiver will be able to read the message content, because of the encryption. The encryption does not only cover the message body, but also the subject and other meta data (e.g., recipient address and sender address).

5.2.3.1 BitMessage P2P Network

BitMessage clients are using a peer to peer network to communicate with each other. A peer to peer network is a type of networking system that does not require a central server. Peers are equally privileged, equipotent participants in the BitMessage Network.

Client- and server functions of the PyBitmessage software are contained in the same application. You cannot disable the server functionality in the software, but you can disable the port forwarding in the local firewall. So, the term "BitMessage server" in this chapter means, that the PyBitmessage core is running and the incoming TCP Port is accessible.

Every BitMessage client, which wants to enter the Network can maintain outgoing connections to 8^3 different BitMessage Server. At the first start of the BitMessage client, the client tries to create the first connection by performing a bootstrap process. Therefore, it tries to connect to 9 hardcoded peers:

- 5.45.99.75:8444
- 75.167.159.54:8444
- 95.165.168.168:8444
- 85.180.139.241:8444
- 158.222.217.190:8080
- 178.62.12.187:8448
- 24.188.198.204:8111
- 109.147.204.113:1195
- 178.11.46.221:8444

Additionally, it attempts to connect to *bootstrap8080.bitmessage.org* and *bootstrap8444.bitmessage.org*, or, in case of using a tor proxy, the onion peer *quzwelsuziwqgpt2.onion*.

After establishing a connection to the first BitMessage peer, the client request connections information to other known peers. This list will be processed sequentially and the client tries to establish connections to other clients until it reaches the

---

[3] https://github.com/Bitmessage/PyBitmessage/blob/v0.6/src/bmconfigparser.py#L16.

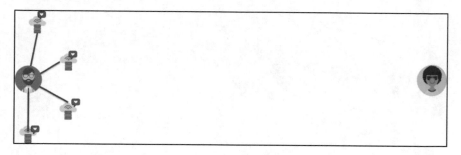

Fig. 5.1 Schematic visualization of a sending process (Step 1)

maximum amount of outgoing connections. The configuration default of the amount of incoming and outgoing connections is 200.[4]

Every single peer repeatedly downloads messages from other peers and puts them into the database. If a new message (not already stored in the database) is received, the peer distributes it to all connected peers. If the received message is already stored in the database, it will be discarded.

Schematic visualization of a sending process:

Step 1: The sender creates an encrypted message ( ). The message is propagated to reach the recipient. Therefore, the sender transmits the message to all connected nodes ( ) (Fig. 5.1).
Step 2: Every single peer, which receives the message tries to decrypt the message. If this failed, it distributes the message to all peers in its connection queue (Fig. 5.2).
Step 3: Sometimes, the recipient's peer will receive the message, If it's part of the network. The network tries to deliver the message to as many nodes as possible until the message's TTL (time to live) is reached (Fig. 5.3).

5.2.3.2   The Message Transfer Process

In this chapter, the object creation and delivery process will be explained by reference to a message object. Other objects (e.g., key exchange or broadcast) have a similar process.

To send a message from a sender to its recipient, the BitMessage client has to be connected to the BitMessage network. Therefore, a bootstrapping process has to be performed.

The concrete message transfer process begins after a user wrote a message to a recipient and activates the send button. In the first step, PyBitmessage is checking, whether the needed public key of the recipient is already available or not. If the key

---

[4] https://github.com/Bitmessage/PyBitmessage/blob/v0.6/src/bmconfigparser.py#L19.

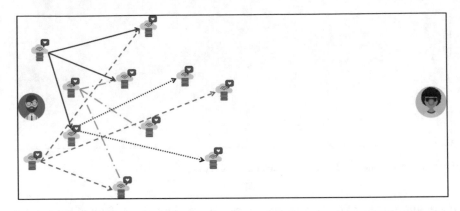

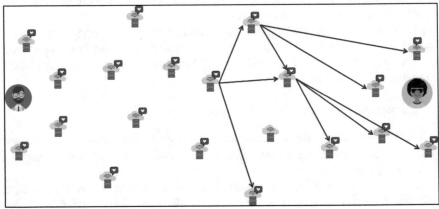

Fig. 5.2 Schematic visualization of a sending process (Step 2)

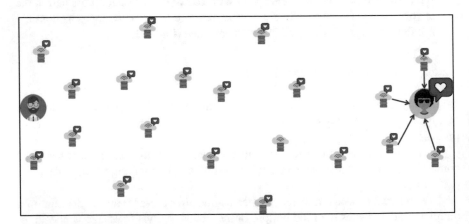

Fig. 5.3 Schematic visualization of a sending process (Step 3)

is unknown, the message transfer process stops at this point and the sending will be queued (Fig. 5.4).

Now PyBitmessage tries to obtain the public key by requesting it from one of the connected peers. Therefore, it sends a request to all connected peers. If these peers have saved the key in their database, it will be returned. Otherwise, they'll send the request to all of their connected peers, and so on. This "chain request" continues, until either a client knows the public key or the TTL value of the public key request will be exceeded.

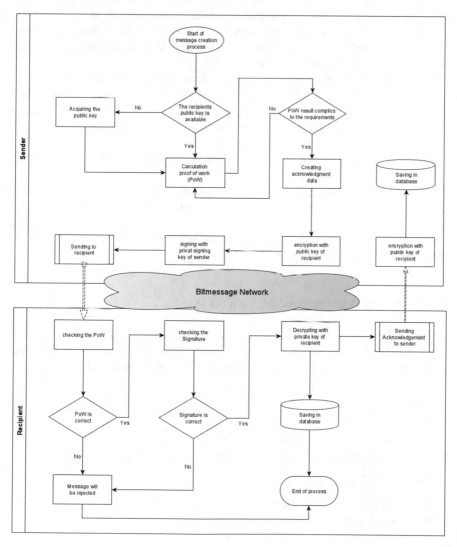

Fig. 5.4 Flowchart of the message transfer process

Because of the circumstance, that the public key request doesn't contain inquirer information, the public key has to be transferred to all participants of the Network, and thus it reaches the inquirer, too.

After the recipients, public key reaches the inquirer, or if a request wasn't necessary because the key was known, the message transfer process can continue.

PyBitmessage has to calculate a Proof of Work (PoW). The result of this computing will be transmitted, together with the encrypted message data to the recipient.

The next step is to create an acknowledgment object. In the BitMessage network, the sender and recipients don't know each other. Especially IP addresses are unknown. Neither the sender has information about the IP address of the recipient nor knows the recipient. So there's no direct way to determine, if a message has reached the recipient, or if it has to be sent again.

Because of this, its designated, that the recipient automatically sends an acknowledgment to the sender, that the message was received. Technically, this acknowledgment object is an empty message object with a computed PoW. The complete acknowledgment object has to be predefined by the sender of the original message. Also, the PoW has to be computed by the sender.

Now, PyBitmessage encrypts and signs the message data and transfers this data object to all connected peers.

After the data object reaches the recipient, the data will be decrypted and the data signature checked. Now the PoW will be checked. If one of these steps fails, the message will be discarded and not saved to the database. Otherwise, the recipient will return the contained acknowledgment packet to the P2P network and save the decrypted message into its database.

5.2.3.3 Calculating the Proof of Work (PoW)

To avoid spamming, the sender has to perform a cryptographic calculation of a proof of work, whose difficulty is set by the recipient. This calculation slows down the sending process, dependent on the calculation difficulty and the used hardware.

The calculation of a PoW is affected by the recipient of the messages and the message size.

The sender defines.

- The *payloadLength* (encrypted message object size)
- The TTL (Number of seconds in between now and the object expiring time)

The recipient defines.

- the *averageProofOfWorkNonceTrialsPerByte* (difficulty–default: 1000)
- the *payloadLengthExtraBytes* (to raise the virtual size of the message)
- A target value is calculated by this formula:

Table 5.1 Unencrypted message data

Size	Description	Data type	Comments
Unencrypted message data			
1+	msg_version	var_int	Message format version
1+	address_version	var_int	Sender's address version number
1+	stream	var_int	Sender's stream number
4	behavior bitfield	uint32_t	Client information
64	public signing key	uchar[]	(prepended with \ × 04)
64	public encryption key	uchar[]	(prepended with \ × 04)
1+	nonce_trials_per_byte	var_int	Proof of work
1+	extra_bytes	var_int	Proof of work
20	destination public key	uchar[]	The ripe hash of the public key of the receiver of the message
1+	encoding	var_int	Message encoding type
1+	message_length	var_int	Message length
Variable	message	uchar[]	The message
1+	ack_length	var_int	Length of the acknowledgment data
variable	ack_data	uchar[]	Acknowledgement data
1+	sig_length	var_int	Length of the signature
Variable	signature	uchar[]	ECDSA signature

$$target = \frac{2^{64}}{nonceTrialsPerByte(payloadLengthExtraBytes) + \frac{TTL(payloadLength+payloadLengthExtraBytes)}{2^{16}}}$$

The sender has to calculate a PoW until the result of the PoW is less than the *target*. The initialHash is the SHA612 hashvalue of the encrypted payload.

```
while trialValue > target:
    nonce = nonce + 1
    resultHash = SHA512(SHA512( nonce || initialHash ))
    trialValue = the first 8 bytes of resultHash
```

5.2.3.4   Structure of a Message Object

Messages in BitMessage are represented by an unencrypted message structure. The structure is encapsulated in an encrypted message structure (Tables 5.1 and 5.2).

The message is a UTF-8 encoded string. It uses 'Subject' and 'Body' sections[5]: 'Subject:' + subject + '\n' + 'Body:' + message.

---

[5] https://github.com/Bitmessage/PyBitmessage/blob/v0.6/src/helper_msgcoding.py#L44.

Table 5.2 Encrypted message object

Size	Description	Data type	Comments
Encrypted message object			
8	Nonce	uint64_t	Nonce used for the Proof Of Work
8	ExpiresTime	uint64_t	The "end of life" time of this object
4	ObjectType	uint32_t	Message Object = 3
1 +	Version	var_int	The object's version
1 +	Stream number	var_int	The stream number
?	ObjectPayload	uchar[]	Encrypted message object

Table 5.3 Server specification

Item	Specification
CPU	Intel Xeon D-1540
RAM	64 GB
Bandwidth	500 Mbit/s
Storage	2 × 2 TB HDD SATA Soft RAID
OS	Ubuntu Server 18.04 "Bionic Beaver" LTS

The acknowledgment data contains a new message object with the proof of work already completed that the receiver of this message can easily send out (Table 5.3).

5.2.4 BitMessage Implementations and Applications

5.2.4.1 PyBitmessage

PyBitmessage is an "all-in-one" communication software to communicate within the BitMessage network. It is the first client, which has implemented the network. It has been published in 2012 at Github[6] under the MIT License. The first Code was committed by Jonathan Warren on 12.11.2012.

The software version is 0.6.3.2 with a release on 13.02.2020 (Fig. 5.5).

5.2.4.2 Bitpost

Bitpost is the second Client (Fig. 5.6), which tries to cover the full BitMessage range of functions. It has been published in 2014 at Github[7] under the MIT License. Different from PyBitmessage, there seems to be no further development.

---

[6] https://github.com/Bitmessage/PyBitmessage.

[7] https://github.com/VoluntaryLabs.

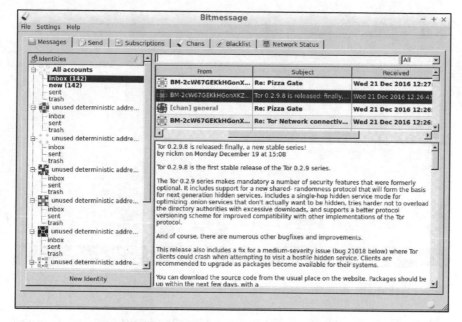

Fig. 5.5 PyBitmessage GUI (inbox)

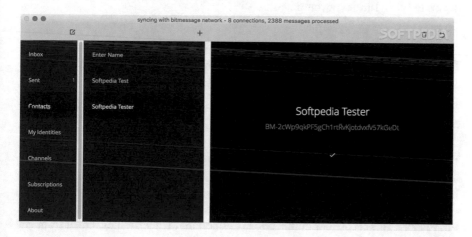

Fig. 5.6 Bitpost GUI

5.2.4.3 Pechkin

Pechin is the only BitMessage client for mobile devices (Fig. 5.7). Additionally it's compatible with Microsoft Windows and Linux. It covers just the messaging part of the BitMessage Network and isn't able to work as a server.

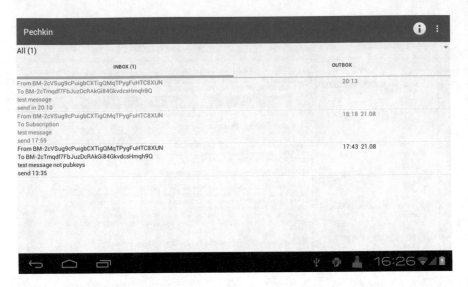

Fig. 5.7 Pechkin GUI

Pechin has been published in 2017 at Sourceforge[8] under the Apache 2.0 Licence. For Android mobile devices, it's available at the Google Playstore.[9] Since 2017 there seems to be no further development.

5.3 Criminal Usage of the BitMessage System and Investigation Challenges

5.3.1 Illegal Transactions

In the last two decades, a lot of criminal shop system or trading forums were coming and going. These pages were reachable in the Clearnet and the darknet (e.g., as TOR hidden service). The operators and users of these services were always dealing with one issue. How can secure communication be established between the customer and the vendor? This secure communication is important to hide an illegal deal from law enforcement and the rivalry.

In the early days, ICQ was a popular communication method. But there was no encryption. Later, encryption addons like Off-The-Record have been used, which added encryption capabilities to some communication protocols. But most of the messaging systems have central servers. The messages may be encrypted, but the identity of the sender and recipient can be revealed.

---

[8] https://sourceforge.net/projects/pechkin/.

[9] https://play.google.com/store/apps/details?id=pro.fenenko.pechkin.

After the development of BitMessage, it was used in the criminal shop systems or trading forums. Especially offenders in child exploitation or child pornography are using this technique.

5.3.2 Blackmailing of Politicians and Celebrities

In 2019 and 2020 a lot of politicians and celebrities received emails from a sender called "Staatsstreichorchester" (coup d'etat—orchestra). In these emails, the sender distributes death threats against politicians, attorneys, and journalists. The writings have a nationalistic and extreme right-wing background [14, 15].

To establish communication, the responsible person offers a BitMessage address as reachability [16].

5.3.3 Command and Control Communication of the Chimera Malware

In 2015, the Chimera ransomware was a ransomware infection. Ransomware refers to threats that take the victim's computer or file hostage and then demand payment to return them. The Chimera Ransomware carries out a basic version of this attack, encrypting the victim's files. This means that, even if the Chimera Ransomware is removed, the encrypted files are not recoverable without the decryption key (which is not supposedly available until the victim pays a hefty ransom using BitCoins) [17].

But added to this feature one more twist that is supposed to put more pressure on the victim. It threatens that in case if the ransom will not be paid, all the stolen files are going to be published, along with the stolen credentials allowing to identify files' owner [18].

When Chimera infects a user it uses an embedded PyBitmessage application to send a Bitmessage to the developer that contains information such as the victim's private key, the victim's hardware ID, and the victim's payment bitcoin address. If the victim is willing to pay, the offender transmits the decryption keys likewise through the BitMessage network [11].

5.3.4 Investigation Issues

Law enforcement authorities have to investigate crime and also identify offenders and their identity. In the scope of IT-related investigations, the IT support measures often comprise of offender identification, lawful interception to enrich investigation information, and forensic data extraction and data analysis.

5.3.4.1 Identification of Communication Participants

The greater part of internet and IT-related crime is committed in the cloak of anonymity. Offenders are using multiple systems to keep their identity hidden. An often used anonymization technique is TOR or I2P, but also BitMessage in reference to Messenger services. Identification measures have a huge stake in the investigation. Often the impenetrability of the anonymization and encryption techniques detains successful investigation.

In the case of BitMessage, the creators of the network are propagating the secure anonymity of participants. This chapter will show, although a decentralized peer-to-peer network is used, there are possibilities to determine the recipient of a BitMessage message.

5.3.4.2 Lawful Interception

Investigations are often divided into a covert and an open phase. Contingent on the case and its quality, the covert phase is containing enrichment of information, which is necessary for a successful investigation. Lawful interception is a possible measure of this phase. Referring to BitMessage, wiretapping of the subjects' internet connection can possibly confirm or deny the results of the IP identification.

5.3.4.3 Data Extraction

A culmination of all investigation measures is to search the subject's apartment and employment. Forensic data backup of the uses IT hardware, and afterward processing and analyzing of the data is mandatory for the success. Because of the multiple kinds of data storage, investigators often struggle while preprocessing the data. Especially, different messenger services are using completely different data storage formats. Thus, a uniform and suitable preparation are compound.

5.4 Adopted Approach

5.4.1 Identification of Recipients

5.4.1.1 BitMessage Original Sender Identification

The network delivery system of Bitmessage is based on flooding. Flooding is used in computer networks routing algorithm in which every incoming packet is sent through every outgoing link except the one it arrived on [19].

If the message delivery process of all delivering servers can be monitored, it is possible to determine, which participant is the originator of a message. The incoming connections of the delivering servers have to be sorted by timestamp if the wanted message is included. The sender of the first determined in time has to be the originator of the message. Therefore, two main issues have to be solved.

The adjustments of the delivering server's time generators have to be coordinated to ensure a correct comparison of the incoming messages by time. The transport of messages inside a computer network can be relatively fast, thereby small inaccuracies in the time systems may result in a false outcome.

The next essential part is the accurate identification of a single message inside the messengers' data stream. Only this unique identification enables the above-mentioned comparison.

The control of all delivering servers is inpossible. But if a client is connected to more than a single delivering server it's not necessary to log the incoming data of all of them. It is enough to control at least one of the connected servers to retrieve a sufficient result.

5.4.1.2   Bitmessage Identification Approach

In general, if sending a message inside the BitMessage network, sooner or later every single BitMessage Client will be a relay for this message. Nearly 100% of all clients, which are connected in the messages TTL, will receive the message. And all clients except the recipient will resend it. The challenge in identifying the message origin is, to identify the first sender of the message.

Every BitMessage client can handle up to eight connections to BitMessage servers. If BitMessage clients operate as a server too, it can handle up to 200 connections (depending on the hardware) including the outgoing connections, by default.

If every client is connected to at least one altered monitoring client, it seems to be possible to determine the originator of a message with a high probability. Therefore two main challenges have to be resolved.

1. Fast packet relaying to our monitoring nodes
2. Identifiability of a single message object

Packet Relaying

Each connection is managed by a data queue. The component, handling the relay or the own data creation puts the data, selected for sending in these queues. Depending on the current queue length and the particular data connection speed, an incoming packet will be relayed at different points of time. PyBitmessage tries always to serve the fastest connected nodes at first (Fig. 5.8).

Let us consider a single message, observed by one monitoring client. Let's also assume, that the originator of a message is directly connected to the monitoring client. We don't know any information about the other clients, which are connected with the originator. And we also don't know the number of clients. But it's very likely,

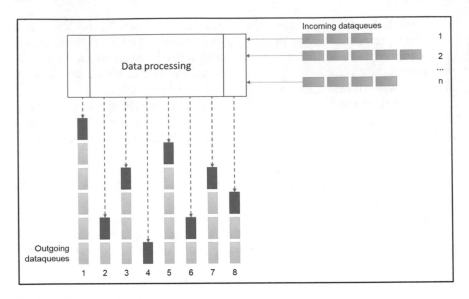

Fig. 5.8 BitMessage data relay

that the originator has not server capabilities. So the maximal number of connected clients is eight ($c_{org} = 8$).

Let us first calculate the probability that the originator sends the message to the monitoring client on one specific order of precedence.

$$p(A) = \frac{1}{c_{org}} = \frac{1}{8}$$

The probability, that the monitoring client receives the message at first is $\frac{1}{8}$.

As mentioned before, we know nothing about the connection quality of the originator connections. So it's possible, that the connection quality (and so the connection speed) between the originator and the monitoring client is worse than the connection quality between the other connected nodes. The monitoring client may receive the message from a third node, although it is connected directly to the originator.

In this case, the presumption, the first received message is sent by the originator is wrong.

Among others, the connection quality between nodes is dependent on multiple values:

- The distance between the terminal connections
- The current usage of the internet connection
- the used hardware
- Firewall or packet filter.

PyBitmessage is ranking the outgoing connections by their quality. It assigns a value between −1.0 and 1.0 to each node. The higher this value is set, the higher is the probability, that the directly connected monitoring client gets the message from the originator at first.

Although we can't affect the originator's internet connection and its hardware, we can choose a fast setup for the monitoring server e.g., fast internet connection, fast CPU and GPU support for the cryptographic functions.

Identifiability of a Single Message

All BitMessage data objects, which are transferred, are heavily encrypted. This also includes information about the sender and the recipient. So the issue is, to identify a data object from a specific origin (e.g., a Message from a specific communication partner).

There are two basic approaches, how we can solve these issues.

The global approach is to monitor all data packets. The monitoring nodes collect information about all message packets, relaying by itself, and save this information in a common database. Likewise, the exact timestamp of receiving will be saved. A hash of the encrypted data packet can be used as an identification feature. If we get a message, whose origin has to be identified, we have to check in the database, from which node one of our monitoring nodes received the message.

Depending on the number of messages and the size of the BitMessage Network, the database would be huge. It also can be a legal issue, if we observe and save all transactions.

A more specific approach is to find a way to identify messages from a specific target. As mentioned in 0, the recipient of a message sends automatically and immediately an acknowledge message back to the sender. This acknowledge-message is predefined and encrypted by the sender. The content of the acknowledge message can be used to track.

Let's explain by using an example: The sender (we) sends a message to a recipient (target). The message is encrypted by the public key of the recipient. Inside the message object, the sender predefincs an acknowledge-message with a unique identifier (e.g., tracking identification) as the message text. This acknowledge message is encrypted by the public key of the sender. The recipient cannot "look" inside the acknowledge-data, but he immediately sends this packet back. The placed monitoring nodes have the sender's private key, so they can decrypt the acknowledge-message. If they note this message, they can save the unique identifier and the timestamp in the common database.

5.4.1.3 Experimental Assembly

To test the hypotheses, a dedicated server with a static internet connection was used.

The datacenter firewall had been configured to forward the ports 22 (SSH access), 8444–8459 (BitMessage Nodes), 80 (HTTP Server). For security reasons, all other ports were closed.

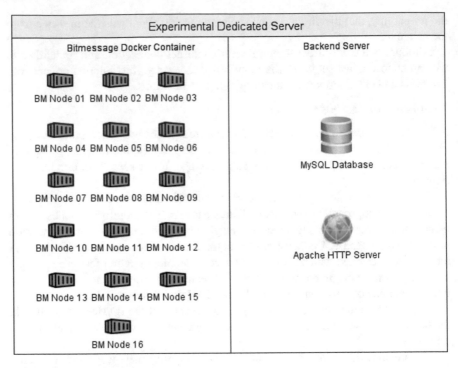

Fig. 5.9 Structure of the experimental server

BitMessage Server

For creating and configuring multiple BitMessage servers, a docker container structure with a Debian derivate as a data image was used (Fig. 5.9).

```
sudo apt update
    catalog update of the APT packet manager

sudo apt install docker.io
    installation of docker

sudo systemctl start docker
    Start of the docker system

sudo systemctl enable docker
    Configuration docker for starting during system boot
```

To use a system inside of the docker container structure, a system image either has to be created or a standard image has to be download. Because the PyBitmessage client doesn't need very special prerequisites, a standard ubuntu-server image was selected.

For downloading and installing a docker image from the docker hub, the "pull" command has to be used (Fig. 5.10).

Fig. 5.10 Download of the Ubuntu Docker image

After starting a ubuntu docker container without any specifications, the necessary prerequisites for PyBitmessage were installed. Considering that, the docker container contains a Ubuntu Linux system, the installation of the prerequisites could easily be done with the apt packet manager.

```
sudo apt-get install python openssl libssl-dev git python-msgpack
python-qt4
```

After this, the container was detached and the changes to a new docker image with the name "ubuntu-bitmessagenode" were committed.

```
docker commit CONTAINERID ubuntu-bitmessagenode
```

In order to handle the container maintenance more efficient, the image "ubuntu-bitmessagenode" doesn't contain the PyBitmessage sourcecode. If parts of the code have to be changed, the codes in all containers have to be altered. So the sourcecode stayed in a shared folder, which was accessible by all docker containers.

Same with the PyBitMessage configuration files. All docker containers have access to a shared folder, containing the PyBitmessage configuration file (key.dat). Because the BitMessage server nodes are running on the same server, they are also using the same IP address. This is the reason, why each container has to listen to different TCP ports. Therefore, the file key.dat contained a placeholder for the port, which had to be replaced during the container start (Fig. 5.11).

Besides the standard configuration, the key.dat file contains the owner's private key, too.

Folder in parent system	Folder in docker container	Description
/root/DockerShare/BitMessageSrc	/root/BM/BitMessageSrc	Modificated PyBitmessage source code
/root/DockerShare/BM2	/root/.config/PyBitmessage	PyBitmessage configuration folder

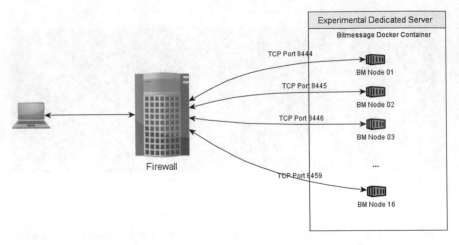

Fig. 5.11 Schematic port forwarding

By using the start script (Fig. 5.12), a single BitMessage Node can be started. Therefore, the script performs the following steps:

1. Starting a Docker container, using the image "ubuntu-bitmessagenode". The container is named "BM_Node" + port number (e.g., BM_Node4888).
2. Copying a new key.dat configuration file into the containers BitMessage configuration folder, by changing the keysVORLAGE.dat file and replacing the port number placeholder "###PORT###"
3. Cleaning up the containers' BitMessage configuration folder.
4. Starting Bitmessage.
5. Starting the container's bash

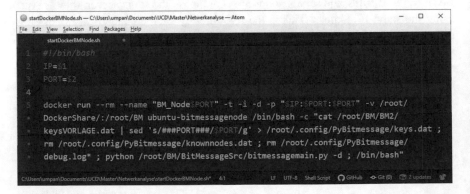

Fig. 5.12 Bash Script to start a Docker BitMessage Node

Fig. 5.13 Bash Script to (re)start all BitMessage Docker Nodes

To (re)start all BitMessage Docker nodes, the following start script was used (Fig. 5.13).

Tracking Draft

The concept of my tracking system contains three parts (Fig. 5.14).

First, there is a customized BitMessage client. This client can be used to send a message to the tracking target. Therefore, it generates a *unique tracking id*, using the current Unix timestamp. Furthermore, it generates the acknowledgment message and inserts into its content the *unique tracking id*.

After sending the message, it sends the tracking id and some information about the target address to the tracking server.

The next part of the tracking system is the BitMessage Nodes. After receiving the tracking message, the target returns the acknowledgment message. If a BitMessage node takes notice of the acknowledgment message, it sends the containing tracking id and the current timestamp to the tracking server.

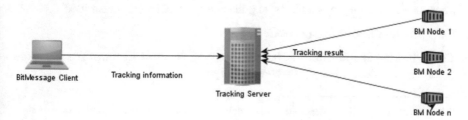

Fig. 5.14 Simple tracking schema

The last part is the tracking server itself. It's a passive database/webserver application. The received information from the BitMessage nodes and the BitMessage clients were collected and processed for the analyst.

To simplify the communication between BitMessage Client and Tracking server as well as BitMessage Nodes and Tracking server, small HTTP GET requests were used.

Code Changes in BitMessage Client

The acknowledgment message can be used to hide information. But the structure of PyBitmessage doesn't make it necessary to alter the process of creating the acknowledgment message. PyBitmessage uses a simple approach to detect, that a relevant acknowledgment message has been received. It doesn't decrypt the message, because the content of the message is irrelevant. It just needs to know, that the ack message has been received. So, after sending, it saves a hash value of the ack message. From then on, every single incoming message will be checked against this hash. This approach is much faster than trying to decrypt the ack message.

The GUI based creating and sending of a message is done in the file./src/bitmessageqt/__init__.py.[10] We changed the code in line 2167.

```
2167 tmp = genAckPayload(streamNumber, stealthLevel)
2168 logger.debug('[TRACKING] '+hexlify(tmp))
2169 id = str(int(time.time()))
2170 ackdata = unhexlify('000000020101eba1') + "FUA_" + id +
     unhexlify('87c33590cad51571557a20b714dc9b2a')
2171 logger.debug('[ZISC] API Send message - stealthLevel: ' +
     str(stealthLevel))
2172 logger.debug('[ZISC] API Send message - toAddress: ' +
     toAddress)
2173 logger.debug('[ZISC] API Send message - fromAddress: ' +
     fromAddress)
2174 logger.debug('[ZISC] API Send message - ackdata: ' +
     hexlify(ackdata[:8]))
2175 logger.debug('[ZISC] API Send message - ackdata: ' +
     hexlify(ackdata))
2176 urllib2.urlopen('http://51.xx.xx.xx/addTracking.php?
     trackingID='+id+'&bmAddress='+toAddress+'&bmInfo='
     +urllib.quote_plus(acct.getLabel(toAddress))).read()
```

In line 2170 the bytes of a former acknowledgement message were taken and the string "FUA_" and the tracking id were inserted.

In line 2176, the URL http://51.xx.xx.xx/addTracking.php was requested and the tracking id, the recipient's address, and the address label were transmitted.

Code Changes in BitMessage Node

The PyBitMessage functions to handle incoming acknowledgement messages are present in the file./src/class_objectProcessor.py.[11]

---

[10] https://github.com/Bitmessage/PyBitmessage/blob/v0.6/src/bitmessageqt/__init__.py#L1986.

[11] https://github.com/Bitmessage/PyBitmessage/blob/v0.6/src/class_objectProcessor.py#L132.

132def checkackdata(self, data, host = "", port = 0, portTCP = 0, selfhost = "", selfport = 0, timestamp = 0):

   ...

```
141   searchID = unhexlify('000000020101eba1')
142    if data[readPosition:readPosition+8] == searchID:
143      id = str(data[readPosition+8+4:readPosition+8+4+10])
144    logger.warning('[TRACKING-RESULT] (' + selfhost + ';' + host
       + ';' + str(port) + ';' +str(timestamp))
145       try:
146         ServerID = selfhost
147         logger.warning('http://51.xx.xx.xx/addLog.php?tracki
ngID=
    '+id+'&serverIP='+ServerID+'&serverPort=0&clientIP='
    +host+'&epochetime='+str(timestamp)+'&clientPort='
    +str(port)+'&clientPortTCP='+str(portTCP))
148         urllib2.urlopen('http://51.xx.xx.xx/addLog.php?tracki
ngID=
    '+id+'&serverIP='+ServerID+'&serverPort=0&clientIP='
    +host+'&epochetime='+str(timestamp)+'&clientPort='
    +str(port)+'&clientPortTCP='+str(portTCP)).read()
149      except:
150      logger.warning('Error:    http://51.xx.xx.xx/addLog.php?
trackingID='
    +id+'&serverIP='+ServerID+'&serverPort=1234&clientIP=
    '+host+'&epochetime='+str(timestamp)+'&clientPort='
    +str(port)+'&clientPortTCP='+str(portTCP))
```

First, the original definition of the *checkackdata* function has been changed. Input variables for source and destination IP addresses and ports and the receiving time were added.

In lines 142 and 143, the script is searching for the unique needle inside the acknowledgment packet bytes.

In line 148, the information is transmitted to the tracking server.

Tracking Server

The main components of the tracking server are a relational database (MariaDB 10.1.44) and an HTTP webserver (Appache/2.4.29).

For receiving and storing the tracking information in the database, two PHP scripts and a PHP database handler script have been used.

- **addTracking.php**

  This script is executed, if a new tracking was created by sending a tracking message. The received GET data is saved inside the database.

  - trackinID: Unique tracking id
  - bmAddress: BitMessage Address
  - bmInfo: Tracking description

- **addLog.php**

  This script is executed, if a monitoring server takes notice of the acknowledgment object, which is part of the tracking system. The received GET data is saved into the database. The IP Addresses will be enriched by geolocation and ISP information.

  - trackingID: Unique tracking id
  - clientIP: IP Address of the connected client
  - clientPort: promoted BitMessage Port of the client
  - cilentPortTCP: real TCP port of the client
  - epochetime: Logging timestamp
  - serverIP: IP Address of the monitoring server
  - serverPort: Port number of the monitoring server

- **db.php**

  This script is the database handler for the tracking server. It handles the database connection and performs the database queries.

  To store the tracking and the log information, a small database schema was created, containing two tables: *Trackings* and *Logs*.

  The *Tracking* table (Fig. 5.15) is containing basic information about the tracking message, recipient, and sending timestamp. The *logs* table (Fig. 5.16) is containing all information about the discovery of every single acknowledgment answer and its timestamp and its origin.

  Both tables have a field "trackingID". This field connects both tables relative to each other.

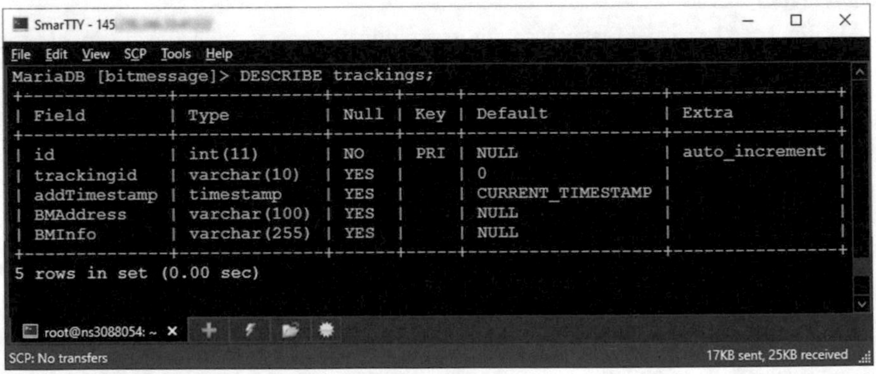

Fig. 5.15 Trackingserver table structure—trackings

```
SmarTTY - 145                                                              —   □   ×
File  Edit  View  SCP  Tools  Help
MariaDB [bitmessage]> DESCRIBE logs;
+---------------------+----------------+------+-----+---------+----------------+
| Field               | Type           | Null | Key | Default | Extra          |
+---------------------+----------------+------+-----+---------+----------------+
| id                  | int(11)        | NO   | PRI | NULL    | auto_increment |
| knownServer         | int(11)        | NO   |     | 0       |                |
| trackingID          | varchar(10)    | NO   |     | NULL    |                |
| serverIP            | varchar(255)   | NO   |     | NULL    |                |
| serverPort          | int(11)        | NO   |     | 0       |                |
| clientIP            | varchar(255)   | NO   |     | NULL    |                |
| clientPort          | int(11)        | NO   |     | NULL    |                |
| clientPortTCP       | int(11)        | YES  |     | NULL    |                |
| clientASN           | int(11)        | NO   |     | NULL    |                |
| clientOrg           | varchar(255)   | NO   |     | NULL    |                |
| clientGeoCountry    | varchar(2)     | NO   |     | NULL    |                |
| clientGeoCity       | varchar(100)   | NO   |     | NULL    |                |
| logtime             | decimal(20,8)  | NO   |     | NULL    |                |
| isVPNServer         | int(11)        | YES  |     | NULL    |                |
| isBMServer          | int(11)        | YES  |     | NULL    |                |
| isTor               | int(11)        | YES  |     | NULL    |                |
| connectionDuration  | int(11)        | YES  |     | 0       |                |
+---------------------+----------------+------+-----+---------+----------------+
17 rows in set (0.00 sec)

MariaDB [bitmessage]> _

root@ns3088054: ~  ×   +   ⚡  ▶  ✳
SCP: No transfers                                            19KB sent, 29KB received
```

Fig. 5.16 Tracking server table structure—logs

5.4.1.4 Network Scale Analysis

The identification approached is based on the fact, that our BitMessage servers are connected with ideally all Bitmessage clients. To approximate, how many own BitMessage servers are necessary, the current scale of the BitMessage network had to be checked.

The unique identification of a single BitMessage client is mandatory for this task. Unfortunately, the network doesn't allow us to get a unique identification ID or string. So another kind of client identification has to be chosen. The only practicable capability for this task is to use the IP address as identifiers. The problem with this selection is that the clients can change their IP addresses and the occurrence of dynamic IP addresses. Currently, there is no way to circumvent this issue.

For guessing, how many own BitMessage servers are necessary, it doesn't require an exact value. It's enough to obtain an approximate value.

The scale of the BitMessage Network can be divided into two different values.

1. The quantity of BitMessage servers
2. The quantity of Bitmessage clients

The Quantity of BitMessage Servers

As mentioned before, the PyBitmessage source code was changed, to write log files in infinite size. Additionally, the hardcoded maximal number of 8 connected Bitmessage servers has been increased. These two changes allowed the counting of connected servers on a daily basis.

When the client finished a connection with a BitMessage server successfully, it inserts the following line into the debug file:

2020–08-10 17:02:41,321—DEBUG—New outgoing connection established: 63.224.149.157.

A simple bash script has been used, to extract the daily quantity of servers. This line, for example, extracts the count for the 10.08.2020:

```
cat debug.log | \
grep "New outgoing connection established" | \
grep "2020-08-10" | \
awk '{print $10}' | sort | uniq | wc -1
```

For this analysis, the BitMessage client was running between 16.07.2020 and 09.08.2020 without a break. The daily results showed (Fig. 5.17), that the range of server count is between a minimum of 18 (18.07.2020) and a maximum of 47 (16.07.2020). On average, there are 27 BitMessage Servers available.

$$\bar{x} = \frac{1}{n} \sum_{i=1}^{n} x_i = 27,26$$

The Quantity of BitMessage Clients

The term "Bitmessage Clients" in this chapter means, that the PyBitmessage core is running, whether the incoming TCP port is open or not.

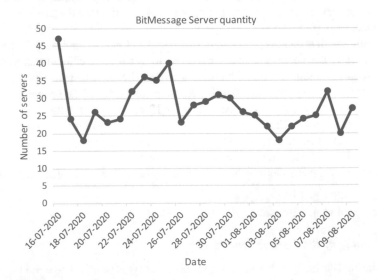

Fig. 5.17 BitMessage server quantity

In the standard configuration of a BitMessage client, it connects to 8 different BitMessage servers. To get an approximate value of the quantity of BitMessage clients, the precondition, that the presumption, every single client is connected to one of my servers has to be created. Currently, there were averaged 27 BitMessage servers in the network. So the injection of 2/3–16 BitMessage servers were appropriate.

Each server is writing a log file. To get an active connection, it's enough to get all log entries, which documents a data transfer. Additionally, I've removed the limit of connected clients, which is normally set to 100, to ensure the servers can connect to as many clients as possible.

2020–07-20   23:06:22,670—DEBUG—89.133.146.180:8444   Requesting   1 objects.

We created a script file BMClientCount.*sh* to extract all entries containing a data transfer by filtering with the term "Requesting" and count them uniquely.

For this analysis, the BitMessage servers were running between 16.07.2020 and 09.08.2020 without a break. The daily results showed (Fig. 5.18), that the range of client count is between a minimum of 1404 (16.07.2020) and a maximum of 1910 (22.07.2020). On average, there are 1702 BitMessage clients, using the network.

$$\bar{x} = \frac{1}{n} \sum_{i=1}^{n} x_i = 1702,08$$

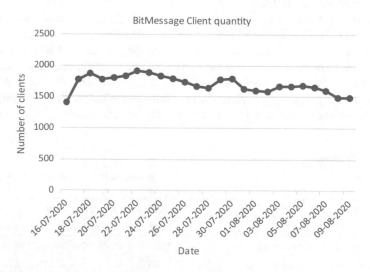

Fig. 5.18 Bitmessage client quantity

5.4.1.5   Auditing

The goal of this approach is to show the possibility to identify unknown users of the BitMessage Network. To audit, the positioned hypotheses, I've conducted some test series.

Test Description

In each test, the circumstances were known (e.g., used client software, IP Address), so the comparison between the tracking results and the expected result was possible. To compare the tests, the following test matrix was used (Table 5.4).

Each target client had to receive 5 messages. The sending process of the first three messages was done, while the target client was online and running. The last two messages were sent, while the target client was offline.

Test Realization

While sending the first test message, we realized, that the key exchange between the sender and the recipient takes a long time. Depending on the client, the exchange took between 30 and 180 min.

After receiving the public keys, the message sending process was (relatively) fast. All sent messages reached the recipient. All clients returned the acknowledge message automatically to the sender. And in all tracking cases, the monitoring servers received the logging information.

The result of the different client software versions differs a little bit from each other. The test series with the windows version of PyBitmessage 0.6.1 (Lastest compiled Version) resulted in all cases the correct IP Address (Table 5.5: Line 6–10). Actually, the expected IP Address was always in the first position of the result list. There seems to be no distinction between the recipients' client is online or offline.

The results of the test series with the PyBitmessage Client 0.6.3.2, running in Linux is quite similar, as the Version 0.6.1. running in Windows. The tracking tests, during the online phases of the client were successful. The expected IP Address was

Table 5.4 Test matrix

Sending date (SD)	Exact date and time when the test message was sent
Target software (TS)	Client software of the target
Target status (Stat)	Target connection status, while sending
Target received message (TRM)	Marker, if the target client received the message
Sender received returned Ack Object (SRA)	Marker, if the sender client received the acknowledgment message
Target IP in tracking result list (TR)	Marker, if the expected IP was in the result list
IP ranking (LP)	Position of the expected IP in the result list

Table 5.5 Tracking test results

#	SD	TS	Stat	TRM	SRA	TR	LP
1	13.8.20 12:55	BitPost 0.9.8.3 (Mac OSX)	Online	Yes	Yes	No	n/a
2	13.8.20 13:06	BitPost 0.9.8.3 (Mac OSX)	Online	Yes	Yes	No	n/a
3	13.8.20 13:10	BitPost 0.9.8.3 (Mac OSX)	Online	Yes	Yes	No	n/a
4	13.8.20 13:24	BitPost 0.9.8.3 (Mac OSX)	Offline	Yes	Yes	No	n/a
5	13.8.20 13:31	BitPost 0.9.8.3 (Mac OSX)	Offline	Yes	Yes	No	n/a
6	13.8.20 13:10	PyBitmessage 0.6.1 (Windows)	Online	Yes	Yes	Yes	1
7	13.8.20 13:20	PyBitmessage 0.6.1 (Windows)	Online	Yes	Yes	Yes	1
8	13.8.20 13:31	PyBitmessage 0.6.1 (Windows)	Online	Yes	Yes	Yes	1
9	13.8.20 13:34	PyBitmessage 0.6.1 (Windows)	Offline	Yes	Yes	Yes	1
10	13.8.20 13:39	PyBitmessage 0.6.1 (Windows)	Offline	Yes	Yes	Yes	1
11	18.8.20 12:21	PyBitmessage 0.6.3.2 (Ubuntu)	Online	Yes	Yes	Yes	1
12	18.8.20 12:33	PyBitmessage 0.6.3.2 (Ubuntu)	Online	Yes	Yes	Yes	1
13	18.8.20 12:34	PyBitmessage 0.6.3.2 (Ubuntu)	Online	Yes	Yes	Yes	1
14	18.8.20 12:36	PyBitmessage 0.6.3.2 (Ubuntu)	Offline	Yes	Yes	No	n/a
15	18.8.20 12:38	PyBitmessage 0.6.3.2 (Ubuntu)ara>	Offline	Yes	Yes	No	n/a
16	18.8.20 12:45	Pechin vA0.1 (Bluestack)	Online	Yes	No	No	n/a
17	18.8.20 13:59	Pechin vA0.1 (Bluestack)	Online	Yes	No	No	n/a
18	18.8.20 13:59	Pechin vA0.1 (Bluestack)	Online	Yes	No	No	n/a
19	18.8.20 14:33	Pechin vA0.1 (Bluestack)	Offline	Yes	No	No	n/a
20	18.8.20 17:51	Pechin vA0.1 (Bluestack)	Offline	Yes	No	No	n/a

always in the first position of the result list (Table 5.5: Line 11–13). However, during
the offline phases, the tracking returned no result (Table 5.5: Line 14–15). In both
cases, the target client wasn't connected to at least one monitoring node. Because
of this, reviewed the monitoring nodes protocol files. The IP Addresses of the target
client were rejected by the monitoring clients because a reconnection attempt was
repeated too often.

The results of the test series with the BitPost 0.9.8.3 Client running under macOS
were different from the other results (Table 5.5: Line 1–5). Although the tracking
mechanism was triggered, the expected IP Address hasn't been determined. The first
entries in the result list were always IP Addresses, possibly belonging to internet
servers. With this insight, the IP Addresses had been checked against the daily list of
TOR Exit Nodes. The result was, that every single IP Address was part of the TOR
Network (Table 5.5).

Concerning the BitPost Github page, the BitPost client always uses the TOR
network to connect to the BitMessage Network. Thus, the determination of a Clearnet
IP Address isn't possible.

The test series with the Pechin vA0.1 client revealed another problem (Table
5.5: Line 16–20). The software development seemed to be aborted and abandoned
extensive parts of the BitMessage protocol were not implemented. The important
implementation of the Bitmessage message acknowledgment system wasn't part of
the software. Thus the tracking resulted in no positive logging results.

5.4.1.6 Conclusion

Generally, the approach of monitoring the BitMessage network to obtain information
about the recipient of a message is working.

As a result, the probability of securely identify the recipient of a Bitmessage
communication is directly dependent on the size of the Bitemessage network and
the amount of inserted manipulated delivering servers and the current size of the
Bitmessage network affords to identify the recipient of a Bitmessage communication
securely, assuming that the recipient does not use a TOR exit node.

5.4.2 Network Surveillance

5.4.2.1 Methodology

To gathering information out of the network traffic of an application, multiple process
steps have to be performed.

Data Capturing

The data capturing is the acquiring and storing of all network traffic data, the software to be monitored is generating. That includes wired or wireless network data as well as other communication devices like Bluetooth.

Data Preparation

To analyze relevant data content, the captured data has to be prepared. This preparation contains the dropping of unnecessary data pieces (e.g., traffic, produced by other software) and preprocessing of the relevant data packets. This preprocessing comprised a recognisability of known data structures. A usual way to perform the data preparation is to use a network analyzer (e.g., wireshark) and a dissector for the expected network protocol. If there's no dissector available, a suitable dissector has to be developed.

Data Analyses

Depending on the resulting data fields of the data preparation, different analyzing methods have to be performed. It has to be analyzed if there are elements in the data field, that contains target-oriented information about the sent message or their meta-data.

5.4.2.2 Experimental Assembly

To determine the operationality of the network traffic, the produced traffic of a Bitmessaege client has to be captured and analyzed. Essentially, the best approach to capture the network traffic of an application is to capture the traffic of the whole computer system and all of its network interfaces. But, because the traffic of the BitMessage network is highly encrypted, and therefore possibly not delimitate to other data, the decision to capture just the traffic of the Bitmessage client application, was made.

Different from other capture software products (eg Tshark, Wireshark) the Microsoft Network Monitor 3.4 has implemented the possibility, to filter by an application. This is, why this (outdated) capture-application was chosen. The captured data packets were saved as PCAP files.

To analyze the PCAP data file, Wireshark 3.2.3 was used, because of its implemented analyzing function.

As BitMessage client, the Win32 compiled PyBitmessage 0.6.1, running under Windows 10 was selected.

The experiment consisted of two phases. First the capturing phase and second the analyzing phase. Before the capturing phase began, all unnecessary applications were stopped. Then the Microsoft Network Monitor (Fig. 5.19) had been started and begun the network capture. After establishing the capture, the PyBitMessage client was started and run until the connection to BitMessage servers were established. Now an external BitMessage client was used to send a tracking message with the

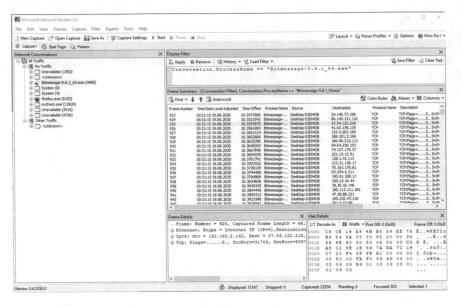

Fig. 5.19 Microsoft network monitor

running client as the recipient. Once the message arrived, the capturing was stopped and the captured data saved.

5.4.2.3 Packet Analysis

The captured data contained network packets between 19.08.2020 10:31:15 and 19.08.2020 10:38:37. The total amount of captured data packets is 12148.

By using the conversation analysis capabilities of Wireshark, the determination of five TCP connections with contained payload data was possible. The residual 1308TCP conversations were just unsuccessful connection attempts, which were irrelevant for this analysis.

To filter the relevant network packet, the following filter string as a *display filter* was used.

```
(ip.addr==175.0.120.133)    ||    (ip.addr==31.170.186.69)    ||
(ip.addr==95.111.7.240) || (ip.addr==87.2.232.217)
```

To reveal the BitMessage packet structure out of the data stream, Wireshark BitMessage LUA dissector of Jesper Borgstrup[12] was used (Fig. 5.20). This dissector recognizes most of the BitMessage object types (version, verack, addr, inv, getdata, getpubkey, pubkey, msg, broadcast object types) Because of the data encryption, the dissector isn't able to display any content.

---

[12] https://github.com/jesperborgstrup/bitmessage-wireshark-dissector.

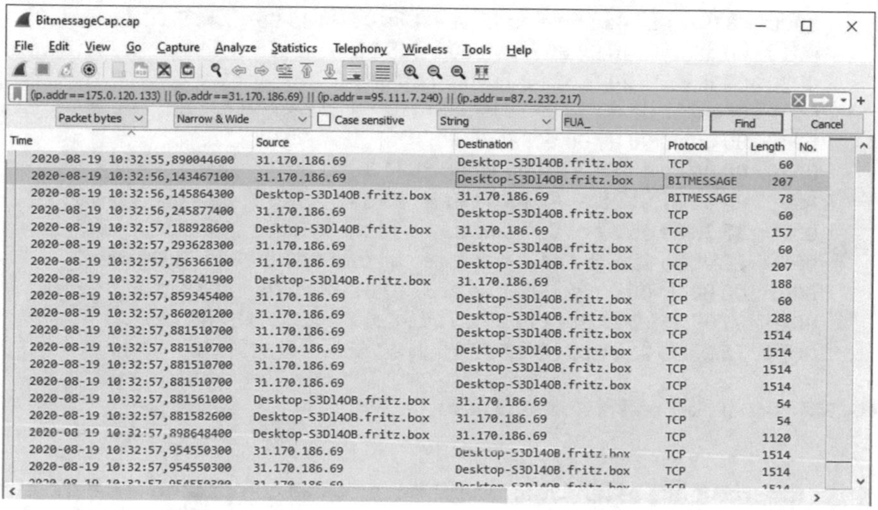

Fig. 5.20 Wireshark packet list

Unsurprisingly, the content of all data packets was encrypted. It seemed, that the encryption approach of the BitMessage network had been correctly implemented in the PyBitmessage client.

Furthermore, the BitMessage dissecter is just able to extract the first packet of a data object. The following data wasn't recognized.

To determine, where the acknowledgment object, used for tracking, can be found inside the data stream, a search for the packets was performed. Therefore, the display filter to filter the Wireshark packet list for the string "FUA_" (see Sect. 5.2.3.3) was set.

```
frame contains "FUA_"
```

The Wireshark's display filter filtered eight data packets, containing the string "FUA_". In the particular packet data, the string and the attached TrackingID is noticeable (Fig. 5.21).

The other packets contained acknowledgment objects of previous trackings. Also objects of trackings, with other target clients.

5.4.2.4 Conclusion

The wiretapping of an internet connection doesn't lead to further knowledge of sent messages, communication partners, or other meta-information sent by a BitMessage Client. Because of the data encryption, a look inside the data payload is not possible. Even if the breaking of the AES-256 is prospective possible, the existence of a specific message packet inside the BitMessage data stream doesn't imply, that the wiretapped BitMessage client is the origin of the message.

```
0000   c8 0e 14 a4 4b b6 54 ee 75 b8 84 8a 08 00 45 00    ....K.T.u.....E.
0010   00 af 0c 4c 40 00 80 06 00 00 c0 a8 02 8e 5f 6f    ...L@........._o
0020   07 f0 7f 25 20 fc b2 92 f8 c0 bd 40 56 b6 50 18    ...% ......@V.P.
0030   06 40 2b 37 00 00 e9 be b4 d9 6f 62 6a 65 63 74    .@+7......object
0040   00 00 00 00 00 00 00 00 00 36 81 af 59 8c 00 00    .........6..Y...
0050   00 00 00 6b 8b f3 00 00 00 00 5f 46 1c d4 00 00    ...k......_F....
0060   00 02 01 01 eb a1 46 55 41 5f 31 35 39 37 38 32    ......FUA_159782
0070   35 38 39 35 87 c3 35 90 ca d5 15 71 55 7a 20 b7    5895..5....qUz .
0080   14 dc 9b 2a e9 be b4 d9 67 65 74 64 61 74 61 00    ...*....getdata.
0090   00 00 00 00 00 00 00 21 59 cf 61 1f 01 18 f3 7d    .......!Y.a....}
00a0   55 07 3c 92 26 3f ea 11 eb 40 e9 a6 af 85 4c 2c    U.<.&?...@....L,
00b0   75 06 99 81 c6 b2 5b 11 79 75 dc e3 d3             u.....[.yu...
```

Fig. 5.21 Data packet—Acknowledgment object

For the scheduling of further investigation measures, information about the user activities of an offender can be useful. If it's known, that a delinquent is using BitMessage, you can use the wiretapping for the information, when he is normally using his computer.

5.4.3 Forensic Information Gathering

5.4.3.1 Methodology

Computer forensics is the application of investigation and analysis techniques to gather and preserve evidence from a particular computing device in a way that is suitable for presentation in a court of law. The goal of computer forensics is to perform a structured investigation while maintaining a documented chain of evidence to find out exactly what happened on a computing device and who was responsible for it [20].

In this particular case, is not a complete computing device or data storage, but a single software product, running on a device. This circumstance doesn't change the computer forensics scope. It simplified the comprehensive approach of forensics.

There are two approaches to gather information on a single software product's behavior. The first approach is a comparison between the data state before the execution of the target software and the state after the execution. The alteration in the states can be used to determine the software behavior. Additional a copy of the memory can be saved before and after the execution to append the changes in memory into the following analyses.

This approach lacks the inclusion of volatile data. The target software possibly overwrites important data changes or temporary data while executing, which can't be determined afterward.

The second approach is to monitor the target software while running. Nearly every current software is based on API calls of the Operation System it is developed for. These API calls include, among others, read and write operations as well as network operations.

Depending on the OS, the target software is running on, there are different software products to perform a behavior recording. If using a Windows Operation system, the Software *Sandboxie*[13] can be used. If using a Linux OS, the software *strace*[14] meets the requirements.

After running the software the data changes have to analyze and categorize. Written data files have to be identified if the data is stored in known data formats or in a proprietary way. To analyze the containing information, the content has to be extracted and classified.

As mentioned above, there are multiple different BitMessage Client, which are capable to communicate inside the BitMessage Network. Besides the two deprecated clients Bitpost and Pechkin, there is PyBitmessage, the Reference client for Bitmessage, written in Python. This client is available as compiled Windows 32 Binary, too.

The following forensic analyses are referring to PyBitmessage, because it's the most used client for communication in the BitMessage network.

5.4.3.2 Observation of the PyBitmessage Affected Data

Often, the forensic analyst hasn't the possibility to test a software product and study its behavior. He has to analyze the given data. Because of the circumstance, that PyBitmessage is an Open Source Software, we can study its behavior by creating and sending prepared messages.

To do this study, a clean lab environment has to be created.

Laboratory Environment

For forensic analyses, a Virtual Machine (VM) is an ideal environment to create and study the behavior of a software product. There are several virtualizing systems to create and run VMs. Because of the reason, that this study just deals with written data on data storage, and takes no account of memory forensics or CPU debugging, the choice of the virtualizer is subsidiary. For simplicity, the decision was to choose Oracle Virtual Box.[15] For these studies, a VM with the following specification has been created.

- 64Bit CPU Virtualization
- 2048 MB Memory
- KVM Paravirtualization

---

[13] https://github.com/sandboxie-plus/Sandboxie/releases.

[14] https://linux.die.net/man/1/strace.

[15] https://www.virtualbox.org/.

- 12 GB SATA Datastrorage (VMDK—Virtual Machine Disk Format)
- OS: Ubuntu (Kubuntu) 18.04.5 Desktop
- Necessary Dependencies for Running the PyBitmessage Software.

Experimental Assembly

By using a Linux Operation System, the surveillance of a running process can easily be performed by using the **strace** tool.

Strace is a diagnostic, debugging, and instructional userspace utility for Linux. It is used to monitor and tamper with interactions between processes and the Linux kernel, which include system calls, signal deliveries, and changes of process state. The operation of strace is made possible by the kernel feature known as ptrace. A system call is a programmatic way a program requests a service from the kernel, and strace is a tool that allows you to trace the layer between user processes and the Linux kernel. System calls are very similar to function calls, which means they accept and work on arguments and return values [21].

Strace writes its output into a text file. The data are in the following format:

```
Structure
System call ( arguments ) = retuned value
```

Example

```
Openat (AT_FDCWD, "/usr/lib/python2.7/copy_reg.x86_64-linux-
    gnu.so",O_RDONLY) = -1 ENOENT
```

- The first word of the line, **Openat**, is the name of a system call being executed. This system call openes a file relative to a directory file descriptor
- The text within the parentheses is the arguments provided to the system call. The File /usr/lib/python2.7/copy_reg.x86_64-linux-gnu.so will be opened read only (**O_RDONLY**). The argument **AT_FDCWD** means, that the path is treated relatively.
- The result after the = sign (which is **-1 ENOENT** in this case) is a value returned by the **Openat** system call. A directory component in pathname does not exist or is a dangling symbolic link.

For the purpose of this analysis, the system calls with the function of disc writing are deeply interesting. To extract these system calls, the regular expression **^.*?(O_APPEND|O_CREAT).*$** has been used.

O_APPEND

The file is opened in append mode. Before each *write*, the file offset is positioned at the end of the file.

O_CREAT

If the file does not exist it will be created.

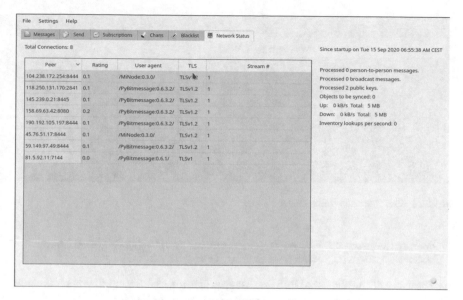

Fig. 5.22 PyBitmessage Forensic Experiment—Execution 01

Experimental Accomplishment

In this experiment, PyBitmessage has been executed three times. Each execution was traced by strace.

Execution 01

See Fig. 5.22.

```
strace -o ~/200915_BitmessageFirstStart.txt \
python bitmessagemain.py
```

In this first execution, the strace recording includes the first (self) configuration of PyBitmessage, the network bootstrap, and first network data synchronization. The traced system calls have been saved into the file *200915_BitmessageFirstStart.txt*.

The system calls has been filtered by the above mentioned regular expression ^.*?(O_APPEND|O_CREAT).*$. Additionaly, all duplicates has been removed.

```
openat(AT_FDCWD, "/dev/null", O_WRONLY|O_CREAT|O_TRUNC, 0666)
= 3
openat(AT_FDCWD, "/home/user/.config/PyBitmessage/debug.log",
O_WRONLY|O_CREAT|O_APPEND, 0666) = 7
openat(AT_FDCWD, "/home/user/.config/PyBitmessage/keys.dat",
O_WRONLY|O_CREAT|O_TRUNC, 0666) = 7
openat(AT_FDCWD, "/home/user/.config/PyBitmessage/knownnodes.
dat", O_WRONLY|O_CREAT|O_TRUNC, 0666) = 13
openat(AT_FDCWD, "/home/user/.config/PyBitmessage/
pybitmessageqt.conf", O_RDWR|O_CREAT|O_CLOEXEC, 0666) = 19
```

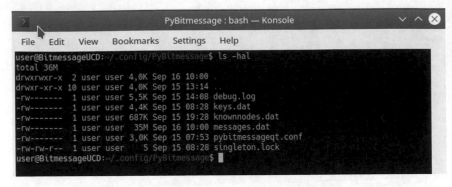

Fig. 5.23 PyBitmessage config folder

```
openat(AT_FDCWD, "/home/user/.config/PyBitmessage/singleton.
lock", O_RDWR|O_CREAT|O_APPEND, 0666) = 12
```

These results implies, that PyBitmessage writes into 5 different files and /dev/null. The Python based complilations from.py to.pyc are not relevant.

1. /home/user/.config/PyBitmessage/debug.log
2. /home/user/.config/PyBitmessage/keys.dat
3. /home/user/.config/PyBitmessage/knownnodes.dat
4. /home/user/.config/PyBitmessage/pybitmessageqt.conf
5. /home/user/.config/PyBitmessage/singleton.lock

A manual check of the folder /home/user/.config/PyBitmessage/ shows, that one additional file has been created: /home/user/.config/PyBitmessage/messages.dat (Fig. 5.23).

This is the main database file of PyBitmessage in the Sqlite3 format. It doesn't appear in the system calls of PyBitmessages, because the Sqlite3 functions were performed by a separated database driver.

Execution 02

```
strace -o ~/ 200915_BitmessageCreateAdresses.txt\
python bitmessagemain.py
```

In this second execution, the strace recording includes the generating of BitMessage Addresses with PyBitmessages. The traced system calls have been saved into the file *200915_ BitmessageCreateAdresses.txt.*

The resulting list of system calls has been filtered in the same way, as mentioned in Execution 01 of the experiment. This execution of PyBitmessage hasn't created additional files.

For test purposes, in this execution phase, two BitMessage Addresses have been created:

- BM-2cT7oFj8qj8DnMu67SKpHGVnxEAnSDHfhA
- BM-2cUx1xwZNkiacjA5iaSecymyfy7sPsmeTu

Execution 03

```
strace -o ~/ 200915_ BitmessageConversation.txt\
python bitmessagemain.py
```

In this third execution, the strace recording includes the sending and receiving of messages with PyBitmessages. The traced system calls have been saved into the file *200915_ BitmessageConversation.txt*.

The resulting list of system calls has been filtered in the same way, as mentioned in Execution 01 and Execution 02 of the experiment. This execution of PyBitmessage hasn't created additional files.

5.4.3.3 PyBitmessage Datafiles

keys.dat

The file keys.dat (Fig. 5.24) contains data of two different issue-areas: basic configuration and Bitmessage address storage. All data is represented in an INI based file format. The basic element is the *key* or *property*. Every key has a name and a value, delimited by an equals sign (=). The name appears to the left of the equals sign. The value, on the right of the equal sign, can contain any character.

The data pairs can be grouped into sections. The section name appears on a line by itself, in square brackets ([and]).

The basic configuration is stored in the ***bitmessagesettings*** section.

For each created BitMessage Address, the keys.dat contains a separate section. These sections are named after the Bitmessage Address.

```
[BM-2cT7oFj8qj8DnMu67SKpHGVnxEAnSDHfhA]
label = Research Address 01
enabled = true
decoy = false
noncetrialsperbyte = 1000
payloadlengthextrabytes = 1000
privsigningkey                                    =
5JbBtuwsDxFFX8ZJAA7RcRjYiZMsKWgwp8dLUJQy1m5TtHB
privencryptionkey                                 =
5Jx41EYRByoY6G2gbhxWgP3xezd1E1bPPzie9A6Gr39m
lastpubkeysendtime = 1600145800
```

Fig. 5.24 Keys.dat—Example address section

knownnodes.dat

The file knownnodes.dat (Fig. 5.25) contains a list of network nodes of the bitmessage network. The data is stored in the JavaScript Object Notation (JSON) data format. Every single network node is represented by an separate JSON Object.

The Object contains the network connection data (host and port), the Bitmessage stream number, and statistical data. The *lastseen* value is in the epoch format.

pybitmessageqt.conf

The file pybitmessageqt.conf contains the configuration data for the user-interface of PyBitmessage. The user-interface of PyBitmessage based on the Qt project. The stored data are for example splitter and button positions.

singleton.lock

The file singleton.lock contains no data at all. It represents, whether PyBitmessage is running, or not.

Fig. 5.25 Example entries in file knownnodes.dat

```
[
  {
    "peer": {
      "host": "170.84.48.19",
      "port": 65519
    },
    "info": {
      "rating": 0.0,
      "self": false,
      "lastseen": 1600143793
    },
    "stream": 1
  },
  {
    "peer": {
      "host": "68.102.185.194",
      "port": 8444
    },
    "info": {
      "rating": -0.1,
      "self": false,
      "lastseen": 1600144396
    },
    "stream": 1
  },
  ...
]
```

```
    2020-09-15 06:55:32,108 - WARNING - No indicator plugin
found
    2020-09-15 06:55:32,109 - WARNING - No notification.message
plugin found
    2020-09-15 06:55:37,113 - WARNING - No notification.sound
plugin found
```

Fig. 5.26 Sample content of debug.log

debug.log

The file debug.log (Fig. 5.26) contains the verbose output of PyBitmessage. Depending on the debug configuration, a complete reflection of the PyBitmessages behavior is stored.

messages.dat

The file messages.dat is the main database of PyBitmessage. It contains messages, foreign public keys and the internal working queue of PyBitmessage.
    The data is stored as SQLite3 Database.

5.4.3.4   PyBitmessage Database Structure

The message.dat (Table 5.6) database file contains 10 different tables and no indices.
    To obtain information about the user communication, three tables have to be regarded.

- Addressbook
- Inbox

Table 5.6 Message.dat

#	Table name	Description
1	Addressbook	Contact list with BitMessage communication partners
2	Blacklist	List of BitMessage addresses, which aren't allowed to send messages to this client
3	Inbox	Incoming Bitmessage messages
4	Inventory	For internale processes—Data source of the process queue of Pybitmessage
5	Objectprocessorqueue	For internale processes—Process queue of PyBitmessage
6	Pubkeys	List of received public keys
7	Sent	Sendet messages
8	Settings	Database settings
9	Subscriptions	Subscription list of Bitmessage channels
10	Whitelist	List of whitelisted Bitmessage Addresses

- Sent

Structure of the database table "addressbook"

#	Field name	Datatype
1	Label	Text
2	Address	Text

Structure of the database table "inbox"

#	Field name	Datatype
1	Msgid	blob
2	Toaddress	text
3	Fromaddress	test
4	Subject	text
5	Received	text
6	Message	text
7	Folder	text
8	Encodingtype	int
9	Read	bool
10	Sighash	blob

Structure of the database table "sent"

#	Field name	Data type
1	Msgid	blob
2	Toaddress	text
3	Toripe	blob
3	Fromaddress	test
4	Subject	text
5	Message	text
6	Ackdata	blob
7	Senttime	int
8	Lastactiontime	int
9	Sleeptill	int
10	Status	text
11	Retrynumber	int
12	Folder	text
13	Encodingtype	int
14	Ttl	int

5.4.3.5   Data Extraction of a PyBitmessage Installation

Extraction of all BitMessage Communication Addresses

The extraction of the used BitMessage entities (BitMessage address) is the main step of gathering the BitMessage usage information of an analyzed system.

To extract all BitMessage communication addresses, created by the PyBitmessage installation, it's enough to explore the keys.dat file. If the analyzed computer system has multiple configured user areas, each of them has to be included.

The file path of the keys.dat file is [user root]/.config/PyBitmessage/keys.dat.

Extraction of Communication

The communication can be revealed by joining the tables *addressbook* with *inbox* and *addressbook* with *sent* of the messages.dat file with this SQL Queries (Figs. 5.27, 5.28, 5.29 and 5.30).

```
SELECT fromaddress,
       toaddress,
       (SELECT address
        FROM   addressbook AS a
        WHERE  i.fromaddress = a.address)  AS  toname,
       subject,
       Datetime(received, 'unixepoch', 'localtime') AS
datetime,
       message
FROM   inbox AS i
ORDER  BY received ASC;
```

Fig. 5.27 Query to acquire data from the inbox table

```
SELECT fromaddress,
       toaddress,
       (SELECT address
        FROM   addressbook AS a
        WHERE  s.toaddress = a.address) AS toname,
       subject,
       Datetime(senttime, 'unixepoch', 'localtime') AS
datetime,
       message,
       status
FROM   sent AS s
ORDER  BY senttime ASC;
```

Fig. 5.28 Query to acquire data from the sent table

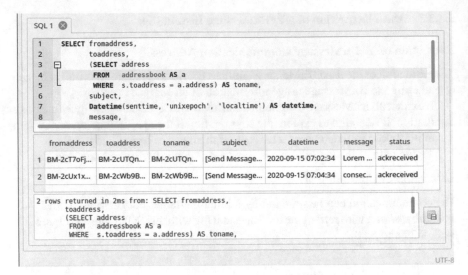

Fig. 5.29 SQLite Browser—Query to acquire data from the inbox table

Fig. 5.30 SQLite Browser—Query to acquire data from the sent table

Recovering of Deleted Messages

Normally, an SQLite3 DB doesn't delete data in the strict sense of the word. If a field or data row has to be deleted, the database just leaps this data. Not before the database engine or the user executes the vacuum command, the leaped data will be finally deleted.

Unfortunately, PyBitmessage performs the vacuum command regularly, so there is (besides the standard forensics measures) no possibility to recover deleted messages.

5.4.3.6 Conclusion

PyBitMessage uses common techniques like SQLite and JSON to store data. This makes the task of gathering information about messenger communication realizable. At least, if the owner of the computer system hasn't deleted any messages.

The main database is stored as SQLite File. No data field is encrypted or obfuscated. Also, the private keys of the used BitMessage account aren't encrypted or otherwise secured.

To extract message information about the saved conversations, just an SQLite capable database management system and a text editor is necessary.

5.5 Summary

This chapter is considered to the structure and technical background of the BitMessage Network and its different clients. Close attention was paid to the encryption capabilities of BitMessage as well as the peer-to-peer packet distribution system.

As experiments, 16 adapted monitoring servers have been created by using Docker container and brought into the Bitmessage network. A separate client to send messages and different clients to receive messages were also connected to the network. The receivers were clients with different messaging products to determine a possible software dependence of the results.

The experiment consisted of a messaging test series. A total of five messages were sent to each recipient, do determine if a correct IP address is detectable.

The results of the experiment confirm the initial assumption, that an identification is possible. Generally, the approach of monitoring the BitMessage network to obtain information about the recipient of a message is working. 16 monitoring servers were inserted into the existing network, containing ordinary 27 servers. So the probability, that an arbitrary client is connected to one of the monitoring servers was very high. Arithmetically, three of the eight connections to BitMessage servers had a monitoring server as a destination. If the number of participants of the BitMessage network rises, the number of monitoring servers has to raise too. In the case, that the increase of the monitoring server doesn't correspond to the increase of the whole BitMessage Network, the presumption of the successful tracking decreases.

The negative results of the test series with alternative clients don't change the overall result. The development of both clients, Pechin vA0.1 and BitPost 0.9.8.3 are abandoned. The use of these clients is immaterial in the BitMessage Network. The far most participants are using the original PyBitmessage reference client.

The test results showed, that the identification approach to identify message recipients can be implemented, regarding the current network scale. The number of network nodes (clients and servers) allows, to insert as much as necessary monitoring clients to fulfill the investigation requirements.

The technical approach is based on the probability, that the target person's BitMessage client is directly connected to a monitoring node. However, this can only be an assumption and is not a fact. In a legal environment, the declaration of investigation results requires provable facts. Handling probabilities is often questionable. This means, that a single identification result isn't sufficient for a conclusive identification result. The more concurrent results can be obtained, the better is the state of facts as a basis for argumentation.

The evidence can be condensed by monitoring a suspect's telecommunications. Here the question was asked which approaches does Bitmessage encryption offer to gather information out of foreign Bitmessage communication?

In the performed experiment, the network traffic of a client connected to the Bitmessage network has been captured while a user was sending and receiving messages. Additionally, the network has also been captured while a tracking measure against the client was performed.

The analysis of the traffic did not provide any indications of a possible decoding option. All content and metadata were fully encrypted in accordance with the Bitmessage specifications. These specifications will not allow decryption in the foreseeable future. Additional data that did not correspond to the specified network protocol were not sent.

The identification approach mentioned above changes the data packets above the encryption level. Small parts of the bit sequence have been changed according to a fixed pattern. This pattern was clearly recognizable in the captured data. If the monitored port were the only port to which the manipulated packet would be delivered, the law enforcement IP identification measures could be determined and thereby confirmed. The distribution system within the Bitmessage network, however, means that every participant receives every data packet. The fact that a manipulated data packet was detected during monitoring does not mean that the monitored connection was the actual origin of the packet. General confirmation of the IP identification measures cannot be made.

The PyBitmessage software does not use any special data storage methods. All data is saved either as a SQLite database or as a JSON object. Messages are stored encrypted and unencrypted as a table in the SQLite database. The messages that concern the examined client are always unencrypted. In addition, the associated keys are referenced.

This chapter showed, that the identification of message recipients in the BitMessage network is possible. The identification approach was tested in a phase, where the overall size of the BitMessage Network relatively small. Roundabout 1700 BitMessage nodes were present on average. If the network size should increase in the future, the experiment of this master chapter has to be repeated and the conclusions accordingly rewritten.

The acknowledgment packets used to identify the recipients are changed. An identifier string was entered. This string is recognizable while inspecting the network packets. If anyone will perform a security audit of the BitMessage network, this string can be noticed and scrutinized. A possible solution is to insert the identifier string, before encryption of the acknowledgment packet.

Thus, the data forensic examination of PyBitmessages wasn't part of this chapter, to complete the knowledge about the BitMessage network, an analysis of the messengers' data storage, and an overview of the memory artifacts, produced by the clients, is desirable.

References

1. Daniel, L., & Daniel, L. (2012). Digital forensics for legal professionals. Syngress.
2. Cents R., & Le-Khac, N. -A. (2020). Towards a new approach to identify WhatsApp messages. In *19th IEEE International Conference on Trust, Security and Privacy in Computing and Communications (IEEE TrustCom-20)*, Guangzhou, China.
3. Wijnberg, D., & Le Khac, N.-A. (2021). Identifying interception possibilities for WhatsApp communication. *Forensic Science International: Digital Investigation, 38*(Supplement), 301132. https://doi.org/10.1016/j.fsidi.2021.301132
4. Young, C., McArdle, R., Le-Khac, N. A., Choo, & K. K. R. (2020). Forensic investigation of Ransomware activities—Part 1. In: N. A. Le-Khac, & K. K. Choo (Eds.), *Cyber and digital forensic investigations. Studies in big data* (Vol. 74). Springer. https://doi.org/10.1007/978-3-030-47131-6_4
5. Scams and Safety, U.S. government (Online). Available: https://www.fbi.gov/scams-and-safety/common-scams-and-crimes/ransomware. Accessed November 06, 2021.
6. Wilkens, A. (2016). Drug trafficking on the Darknet: Raids against four suspects. Heise Online.
7. Senker, C. (2016). *Cybercrime and the darknet*. Arcturus Publishing Limited.
8. Dion-Schwarz, C., Manheim, D., & Johnston, P., Terrorist use of cryptocurrencies.
9. Cimpanu, C. (2021). Scarab Ransomware pushed via massive spam campaign. Bleepingcomputer, 11 23 2017. (Online). Available: https://www.bleepingcomputer.com/news/security/scarab-ransomware-pushed-via-massive-spam-campaign/. Accessed November 21, 2021.
10. Warren, J. (2012). *Bitmessage: a peer-to-peer message authentication and delivery system.* (Online). Available: https:\\bitmessage.org\bitmessage.pdf. Accessed November 06, 2021.
11. Abrams, L. (2015). *Chimera Ransomware uses a peer-to-peer decryption service.* (Online). Available: https://www.bleepingcomputer.com/news/security/chimera-ransomware-uses-a-peer-to-peer-decryption-service/
12. Dworkin, M. (2001). NIST, 26 11 2001. (Online). Available: https://www.nist.gov/publications/advanced-encryption-standard-aes. Accessed November 16, 2021.
13. Paar, C., & Pelzl, J. (2010). *Understanding cryptography.* Springer
14. Zeit Online. (2020). Staatsstreichorchester soll neue Morddrohungen verschickt haben. (Online). Available: https://www.zeit.de/politik/deutschland/2020-05/rechtsextremismus-drohbriefe-staatsstreichorchester
15. F. Staatsstreichorchester kündigt Anschläge an, colorful-germany.de, 07 20 2019. [Online]. Available: https://colorful-germany.de/staatsstreichorchester-kuendigt-anschlaege-an/. Accessed August 16, 2021.
16. Anonymous. (2019). staatsstreichorchester_mails_2. (Online). Available: https://doxbin.org/upload/staatsstreichorchestermails2. Accessed November 16, 2021.
17. EnigmaSoft. (2021). Chimera Ransomware, EnigmaSoft, (Online). Available: https://www.enigmasoftware.com/chimeraransomware-removal/. Accessed August 16, 2021.
18. Malwarebytes. (2015). Inside Chimera Ransomware—The first 'doxingware' in wild," Malwarebytes. (Online). Available: https://blog.malwarebytes.com/threat-analysis/2015/12/inside-chimera-ransomware-the-first-doxingware-in-wild/. Accessed November 16, 2021.
19. A. S. Tanenbaum, *Computer networks* (5th ed.) Pearson Education.

20. M. Rouse. "TechTarget," 05 2015. (Online). Available: https://searchsecurity.techtarget.com/ definition/computer-forensics. Accessed January 07, 2022.
21. G. Kamathe. "Understanding system calls on Linux with strace," 25 10 2019. (Online). Available: https://opensource.com/article/19/10/strace. Accessed January 16, 2022.

Chapter 6
Database Forensics for Analyzing Data Loss in Delayed Extraction Cases

Katherine Moser, Kim-Kwang Raymond Choo⊙, and Nhien-An Le-Khac⊙

6.1 Introduction

Most within law enforcement would agree that cellular devices often play a critical evidentiary role in criminal investigations. This is largely due to the fact that most people today, use a cell phone in some capacity. According to a 2018 report published by the Canadian Radio-television and Telecommunications Commission, there were 31.7 million mobile subscribers in Canada by the end of 2017 [1]. With an estimated population of 36.7 million that same year [2], as a Country, we are fast approaching a 1:1 cell phone to person ratio.

For a digital forensic examiner working in a law enforcement digital forensics lab, extracting and analysing cell phone data is a common request for service [3, 4]. Having said that, there are many factors that will determine whether or not the data on a cellular device can be accessed and extracted. These factors can be device specific, such as the make, model, operating system (OS) or dependant on the device support of the lab's forensic hardware and software tools.

One element that has influenced a digital forensic examiner's ability to extract and analyze phone data is device security. More specifically, the use of different types of passwords and encryption. One highly publicized case that brought these very challenges to the forefront, was the 2015 terrorist shooting in San Bernadino, California. During the investigation, the Federal Bureau of Investigation (FBI) recovered an Apple iPhone from one of the suspects. The iPhone was password protected, which prevented the FBI from accessing the device. After a lengthy court battle with Apple Inc. (to no avail), the FBI paid a private third-party company to unlock the device for them [5].

Since the San Bernadino case made headlines, commercial digital forensic tools have emerged that are available to law enforcement only. These tools provide law enforcement organizations with the ability to have an "in-house" solution that can defeat or overcome the same data protections that restricted the FBI in the San Bernadino file. Another advantage of these advanced forensic tools, is the amount of

© The Author(s), under exclusive license to Springer Nature Switzerland AG 2022
N.-A. Le-Khac and K.-K. R. Choo, *A Practical Hands-on Approach to Database Forensics*, Studies in Big Data 116, https://doi.org/10.1007/978-3-031-16127-8_6

data they can extract. Rather than simply performing a backup of a user's data, these tools can extract a partial or full file system from an iPhone, giving law enforcement much more relevant data to use in criminal investigations. While these advanced tools can access and extract data from iPhones that might otherwise be inaccessible, they are not a one size fits all solution and are very expensive, which places them well outside of budgetary reach for many law enforcement digital forensic labs.

In general terms, these advanced forensic tools use two different methods in order access and extract data from a locked iPhone. The first method is a brute-forcing technique (which means guessing the password until the correct one is found) and the second is a bypass method which works around the passcode altogether. The specific method that a digital forensic examiner can use is dependent on a number of factors including the iPhone's model, Operating System version and state. A device's state is very important to note, because the ability to use the bypass method is contingent on the iPhone's state and a specific set of circumstances being met. In order to access and extract the iPhone's data, the device must remain powered on and the password must have been entered at least once since the device was powered on or rebooted. Within the digital forensic community, this condition is referred to as an 'After-First-Unlock' (AFU) or Hot state.

One major advantage of being able to bypass an iPhones passcode is because iPhones utilize security measures specifically designed to combat brute-force attacks. The ability to enter a password at an iPhone's lock screen is purposely slowed down to only allow one attempt every 80 ms [6]. Additionally, after an incorrect passcode has been entered a certain number of times, additional incremental time delays are introduced before additional attempts can be tried. These additional incremental time delays are shown in the Apple's Platform Security Guide [6].

To put this into context, a four-digit PIN that utilizes the numbers 0 through 9 only has 10,000 possible combinations. Increasing the length to 6-digits adds an additional 990,000 possible combinations. Depending on the passcode or password's length (4-digit, 6-digit, 10-digit) and complexity (e.g. all numbers or alphanumeric) the number of potential combinations continues to increases. With the brute-forcing delays outlined above, and depending on the passcode or password in use, successfully brute-forcing an iPhone could take anywhere from a few minutes or hours to months, years or even decades. For law enforcement, having to wait months or years let alone decades to access potentially relevant evidence on a seized iPhone is not feasible. In addition, the 2016 Supreme Court of Canada ruling in R versus Jordan, imposed timelines for the trial of court cases to avoid breaching the rights of the accused, which were established at 18 months for cases before the provincial court and 30 months for those before the superior court [7].

As of today (August 19, 2021) newer iPhones running more recent versions of the Apple operating system (iOS) can only be accessed using the bypass method as the brute-force method is not supported. With the introduction of this new AFU of Hot state criteria, a new set of challenges when seizing a phone exist that must be considered.

In the past, the best practice for seizing electronic devices was to simply power them off. With cellular phones, doing so ensured three things:

(1)   The device was isolated from the network (cellular and internet) in order to cease the sending and receiving of all communications, especially those designed to remotely wipe all the data on the device;

(2)   There was no need to consider how much life was left in the phone's battery; and

(3)   The device's powered off state did not impact the investigative process, specifically the level of urgency in:

    a.   Obtaining judicial authorizations to search the device (if not already in place);

    b.   Transferring the seized device to the digital forensics lab; and

    c.   When the extraction of data from the seized phone was conducted.

Now that advanced forensic tools require certain phones to remain powered on in order to access and extract their data, previous best practices of powering off a Phone is no longer a viable option. Other techniques must now be used in order to ensure the phone's network isolation, e.g. placing the phone in airplane mode, removing the SIM card or using a Faraday bag. A new consideration introduced by the need for keeping a device powered on is the phone's battery life. Ensuring that a phone does not power off as a result of its battery drain could require access to and the use of charging cables and adapters and access to an electrical power source such as an electric outlet or battery pack.

A phone that must remain in an AFU state means that the Operating System is still running. Therefore the phone's battery will continue to deplete unless connected to a charger, and changes to the device's data are still occurring. For example, a password protected device that is seized and isolated from the network has an alarm set to go off at 06:00 a.m. every day. As long as the device remains powered on, this alarm will continue go off daily at 06:00 a.m. because the device remains powered on and the operating system is still running. The question then becomes, by virtue of the Operating System continuing to run, does the iPhone then become a volatile container for the data it contains? Even if a seized device gets to the digital forensics lab at the earliest possible opportunity, can delays caused by the investigative process (i.e. lab backlogs or the need to first obtain a search warrant before the phone's data can be extracted) cause the loss of potentially relevant data?

In an effort to protect evidence that might be of evidentiary value to an investigation, the proposed research aims to demonstrate that device data is lost when a seized iPhone must be left powered on in order to be extracted, and extraction delays occur as a result of the investigative process. The research will determine if the longer the waiting period extends before extracting the data from a live iPhone, so will the number of artefacts that will be lost on the device.

Within the digital forensic community, there appears to be a working level knowledge or understanding that data on Apple devices can be deleted. This knowledge or understanding is often learned through work experience, training or asking others within the digital forensics community for advice. However; the shared knowledge is often specific to known data retention periods of certain applications or device settings, such as the knowledgeC.db database only storing data for approximately

22–26 days. While this type of information is very useful to a digital forensic examiner, it does not address overall data loss, and it does not address loss specifically related to delaying a device's extraction.

For most law enforcement digital forensic examiners, the ability to conduct the type of research proposed in this paper would not be possible due to the high costs associated with purchasing advanced forensic tools. Additionally, operational requirements would not only take priority for using those tools, but also an examiner's time.

The objective of this chapter is to provide the digital forensic community, and larger law enforcement community as a whole, a research based answer to the question, is data lost when the extraction of a seized iPhone is delayed. This chapter aims to:

- Contribute to digital forensic process best practices in dealing with a seized iPhone;
- Provide research results that could form part of an exigent circumstances framework; and
- Enable better outcomes for criminal investigations.

6.2   Background

6.2.1   iOS SQLite Databases

Most iOS applications store data in a SQLite database, which is "an in-process library that implements a self-contained, serverless, zero-configuration, transactional SQL database engine" [8]. Since these databases are self-contained, they are usually comprised of a single main database, with additional temporary files that store information when a change to the database (called a transaction) is being made [9]. The two most common temporary files found with iOS SQLite databases are the shared-memory file and write-ahead log. This is because the databases run in what is known as WAL mode. The shared-memory file (named for the database with "-shm" added to the end) provides shared memory that the write ahead log uses as an index [10]. The write-ahead log (named for the database with "-wal" added to the end) keeps a record of changes that have been made, but not yet written to the main database file.

6.2.2   SQLite Vacuuming

When data is deleted from a SQLite database, the deletion does not necessarily occur right away. Rather, the area that contained the deleted data is marked or flagged as 'available space' to be used by the database to store new data when it arrives. As a result, data can become fragmented and cause the database to grow in size unnecessarily. This is where the VACUUM command comes into play. When executed, the VACUUM command cleans up any 'available space' areas (in the main database file only) by rebuilding the database file and repacking the data inside the database so it can occupy the smallest amount of required space [11].

In order to keep a SQLite database running optimally, the VACUUM command can be set to run automatically, referred to as 'auto_vacuum = FULL' mode [11]. Running in full mode, a SQLite database uses the 'available space' once data is deleted, which reduces the overall size of the database without the need to run a separate VACUUM command to rebuild the entire database. The auto-vacuum mode has three settings in total, NONE (0), FULL (1) or INCREMENTAL (2) [12].

In order to set the vacuum status of a SQL database to a particular mode, a statement (called a PRAGMA statement) is used. A PRAGMA statement is specific to SQLite, and is used to modify how a database operates [12]. When a database is set to '2' or incremental mode, auto-vacuuming will only occur once a pragma statement with 'incremental_vacuum' is used [12].

6.3   Methodology

In the last number of years, law enforcement digital forensic labs have seen a shift in the electronic evidence they receive. More and more, requests for analysis centre around portable electronic devices, and in particular mobile phones. The introduction of 'in lab' advanced forensic tools has greatly diminished the need to send locked devices away to third party companies for assistance for those labs fortunate enough to have such a solution available. The support offered by these digital forensic tools, for some devices, is contingent on the device remaining in a powered on AFU or Hot state, the question arose on what happens to the data (if anything) on the seized device while it is waiting to be extracted.

In order to be able to answer this question, the proposed research necessitated a law enforcement context. To have research data that would better inform law enforcement on what happens to the data on an iPhone when the extraction is delayed, all aspects of the research needed to simulate a criminal investigation involving the seizure and data extraction of a smartphone. With this idea in mind, the first step was to select test devices that were capable of having their data extracted while in an AFU or Hot state. Equally important, was each device already being populated with user data unknown to the researcher (just as the user data would be unknown to the law enforcement digital forensics analyst). In order to have a similar yet diverse amount of research data to work with, five phones with different chipset architectures and OS versions were selected. One other important aspect was knowledge of each test phone's PIN number or passcode. The purpose of having this information was to ensure that the test devices were in an AFU state in order to start the extraction phase.

In selecting the digital forensic hardware and software to be used for the extraction and analysis portion of the research the main criteria used was that the forensic tools had to be commonly found in (and currently used by) law enforcement digital forensic labs.

Although the selection of devices and forensic tools were listed like steps, they were conducted simultaneously since they are co-dependent. This avoided selecting a test device that:

a.   Could not be extracted in AFU mode;
b.   Was not supported by the digital forensic tools; or
c.   Was supported but only by a forensic tool not available to the researcher.

Once the devices and forensic tools were chosen, a timeline as well as the number of extraction to conduct on each device needed to be determined. In deciding when the devices should be extracted, consideration was given to using timeframes that the researcher believed would represent realistic delays that could be faced during a criminal investigation. With that in mind, the number of extractions performed needed to allow for comparable data sets that could also show a progression of data loss. Therefore, each device was selected to be extracted four times at an interval of 2 h, 24 h, 72 h and 7 days post device seizure.

At the beginning of the extraction phase, each test device started in a powered off state. To avoid data changes or remote wipe signals received through cellular and data network connectivity, each test device needed to ensure connectivity was disabled. This was ensured through removal of SIM cards (if present) before powering on the text devices and placing the device in airplane mode once it was powered on. After powering on each device and disabling network connectivity, each phone would need to be placed into its AFU or Hot by entering the passcode into the lock screen. Once the device was confirmed to be unlocked, it could subsequently be locked by pressing the home button or power button. With the phone now in the required state for extraction, a simulated device seizure time was selected and all interaction with the phone would cease with the exception of connecting it to a charge cable or forensic tool for extraction. The other two considerations for the extraction phase of the research were how to ensure the test phones did not lose their AFU or Hot state and device continuity.

In order to maintain the charge level of each device's battery in order to keep it in its powered on state, when not being extracted, the test devices were connected to a charging cable. To minimize potential risks to the research being conducted, e.g. power outages, additional measures like the use of a Universal Power Supply (UPS) were implemented. To ensure the continuity of each test device, the charging cables, Universal power supply and test phones were kept in a secure room only accessible by the researcher. Keeping the devices in a secure room was meant to simulate the use of a digital evidence room or locker.

Once all of the extractions were completed, a comparison analysis on each device's four extractions would need to be performed using the forensic analysis software. With the sheer number of applications and the amount of data a single device can contain, coupled with the fact that the data on each test device would be unknown to the researcher, parameters would need to be implemented to narrow the focus of the analysis. In order to find some common ground across all five devices, the scope of the research would be narrowed to focus on native OS applications only.

Any losses of data identified through the comparison analysis would require further investigation in order to verify that the data was removed. Therefore, identification and review of the native applications or file system locations that stored the potentially lost data would need to be examined. To further support the loss of data, and any findings from the file system and application review, a timeline and iOS log analysis would also be conducted to potentially establish a cause as well as further confirm any loss of data.

6.4   Database Analyzing of iOS

In order to establish the scope and approach to the research presented in this paper, consideration was given to five main areas. The first involved determining which phones to include in the research. Once this was established, selecting the forensic hardware and software to be used to extract and analyze the iPhone data was considered. The third area of focus was to determine an appropriate extraction timeline that would simulate delays that could reasonably exist as a result of the investigative process. Next, parameters for which artifacts to focus on were established. Lastly, criteria was established for how the analysis would be conducted and what metrics would be compared in order to establish if any data was lost and how. Each of these five areas are discussed in greater detail below.

6.4.1   iPhones for Conducting the Research

The ability to purchase multiple new phones for conducting this research fell outside the cost neutral budget for this research. In addition, newly purchased devices would not be populated with user data, thus requiring the researcher to ensure the new devices had mobile subscriptions (which is an additional cost) and to populate the devices with user data. Since the approach to this research is to simulate a criminal investigation involving the extraction of data from a seized phone, the data on each test device should not be known to the researcher, just as it would not be known to a digital forensic examiner working in a law enforcement digital forensics lab. With these factors in mind, previously forfeit devices (already populated with user data unknown to the researcher) from a law enforcement digital forensics lab were used.

In order to have a cross section of devices (that have different internal architectures) and OS versions, the research required several test iPhones each running a different version of the iOS. The devices also needed to be protected by a known passcode (preferably already in use). This is because as soon as a user sets a passcode on their iPhone, data protection will be turned on automatically [6]. Knowledge of each iPhone's passcode is strictly for the purpose of ensuring the device is in an AFU state as required by the forensic tool. It will not be used to gain access to the iPhone for the purpose of accessing and extracting the device's data.

With these parameters in place, five iPhones met the desired criteria to be used as test devices. The devices included:

- iPhone X (model A1901) running iOS 14.3 with an A11 chipset and 6 digit PIN;
- iPhone 5s (model A1533) running iOS 12.5.3 with an A7 chipset and 4 digit PIN;
- iPhone SE (model A1723) running iOS 14.6 with an A9 chipset and 6 digit PIN;
- iPhone 7 (model A1778) running iOS 10.2 with an A10 chipset and 4 digit PIN; and
- iPhone 7+ (model A1784) running iOS 11.0.3 with an A10 chipset and 6 digit PIN.

Of the five test devices in the list above, it should be noted that the iPhone SE, iPhone 7 and iPhone 7+ had not been powered on or connected to the cellular network in some time. As such, these devices boot times are not current.

6.4.2 Platforms and Forensic Tools

For the purposes of selecting the forensic hardware and software to use for conducting the research outlined in this paper, the following criteria was established:

The forensic tools need to:

- Be available to be used by the researcher; and
- Currently be in law enforcement digital forensic labs; and
- Support AFU data extractions from the test iPhones; or
- Be able to parse the iPhone data extractions for analysis.

Applying this criteria resulted in only one advanced forensic tool option to perform the required data extractions. Acknowledging that being able to extract each iPhone using at least two different forensic tools would assist in validating the data extraction results, the lack of other tool options will be a noted limitation for this research. In addition, due to non-disclosure agreements, and the fact that the focus of this research is not on the forensic tools themselves, the tools will not be named.

In choosing a forensic tool to parse the data extractions and conduct analysis, based on the set criteria, two forensic software programs were available to the researcher, both made by the company "Cellebrite, Inc." The two forensic analysis programs are Physical Analyzer (PA) and Inspector (formerly known as Blacklight made by BlackBag Technologies Inc.). At the onset of the research, both were the most recent versions of the software, being 7.45.0.92 for PA, and 10.3 for Inspector.

With the forensic tools selected, the only other criteria required was to establish parameters for how the tools would be used to ensure consistency in their application and the results they produced. Therefore, all forensic tools (including software/firmware versions and settings/configurations) will need to remain the same throughout the research to ensure that the results are not tainted from using different versions or settings of the forensic tools.

Non-forensic Tools

In order to conduct the research, a laptop running Microsoft Windows 10 Pro, version 20H2, OS Build 19042.1110, Windows Feature Experience Pack 120.2212.3530.0 will be used. In addition, a free to use software program called DB Browser for SQLite version 3.12.2 (available at http://sqlitebrowser.org/) will be used to assist in conducting comparison analysis of SQLite databases.

6.4.3 Extraction Phase and Timeline

The extraction phase will start with each iPhone powered off with no SIM card inserted into the device. Each iPhone will be powered on and unlocked by entering the passcode on the lock screen to place it into an AFU of Hot state. Airplane mode will be enabled on each device and Bluetooth and Wi-Fi connections will be disabled (if not already done so). Subsequently the device's screen lock will be enabled and the lock state verified. The device will then be connected to a charging cable connected to a UPS unit and wall outlet in a secure room only accessible to the researcher.

These specific measures are being implemented to:

- Simulate how a seized iPhone would likely be treated in a law enforcement digital forensics lab;
- Ensure the device cannot connect to a cellular or data network which could alter or erase the phone's data;
- Maintain both the battery level and AFU or Hot state of each iPhone; and
- Mitigate risks that could jeopardize the research process or alter the results, such as power outages or someone other than the researcher interacting with the test devices.

With the idea of simulating a criminal investigation involving the seizure and extraction of data in mind, once each device is powered on and connected to a charge cable, a specific time will need to be selected which is meant to simulate the time of device seizure. From this time forward, user interaction with the device (other than for the purposes of connecting the device to the extraction tool or charge cable) will cease.

Once the seizure of the device has taken place, four separate data extractions will be performed on each of the locked iPhones. The extraction intervals for each iPhone will use the following timeline, and every effort will be made to ensure all data extractions have strict adherence to this timeline:

- Two hours from the identified seizure time;
- 1 day (24 h) from the identified seizure date and time;
- 3 days (72 h) from the identified seizure date and time; and
- 1 week (7 days) from the identified seizure time.

The first extraction at 2 h post seizure is designed to simulate no extraction delays occurring as a result of the investigative process. The latter three extraction time-frames are purposed to imitate three different periods of delay that could be faced as a result of delays in the investigative process. Between each extraction, the iPhone will be placed back on the same charge cable connected to the Universal Power Supply unit and wall outlet in the same secure location.

One limitation of the timed and consecutive nature of the proposed device extractions, is that any one extraction will not be able to be replicated without starting the whole cycle of device seizure and data extraction over.

6.4.4 Artifacts of Interest

Mobile device extractions can produce an enormous amount of data for a digital forensic examiner to analyze. With the methodology of this research requiring 20 iPhone extractions, parameters needed to be set to streamline what the analysis would focus on to ensure the scope did not become too large.

Normally, a law enforcement digital forensic examiner's analysis is guided by the judicial authorization in place and further streamlined by any specific items of interest or evidentiary value requested by the lead Investigator. With this in mind, the researcher's analysis will focus on those artifacts most commonly sought in an investigation, specifically:

- Electronic communications (text messages, instant messages, chats, and call logs);
- Media files (pictures and videos);
- System logs; and
- Any corresponding databases that house the aforementioned data.

With that being said, one type of electronic communication that will not be included in the research are emails, which is purposeful. The reason for this exclusion is because emails stored in the Apple Mail application are not extracted by the advanced forensic tool when performing an AFU or Hot extraction.

One other component that would require the researcher's analysis to be further streamlined is the need to have consistent and comparable data for analysis, while still being able to answer whether or not data is lost when an iPhone extraction is delayed. The artifact types of interest listed above could be produced by hundreds if not thousands of different iPhone applications. Coupling this with the fact that the data contained on each of the test iPhones is unknown to the researcher, and the Apple App Store having 1.8 million different apps available worldwide as of July 2021 [13], the best solution is to analyze only those applications and databases that are installed as part of the iOS.

6.4.5   Analysis Approaches

As stated previously, consistency in how the forensic analysis tools are configured and used will be key to this stage of the research. Just as a digital forensic examiner does not want to report incorrect findings, it will be important to this research to ensure that any discrepancies found when comparing the data extractions from the same device against each other, are not influenced or caused by the researcher.

6.4.5.1   Comparison Analysis

Each of the device's four extractions will be parsed and analyzed using the commercially available forensic tools identified in Sect. 6.4.2. In an effort to answer the problem statement, a comparison analysis of each iPhone's data extractions will be conducted. Two quantitative data sets will be the focus of this comparison:

1. The size of each data extraction; and
2. The total number of each artifact of interest.

These quantitative data comparisons will be reported in tables using a color scheme to emphasize noted changes in size and number. Increases are emphasized in blue, decreases are emphasized in yellow and those that saw both an increase and decrease are emphasized in green. To provide structure to this comparison, the outline depicted in Fig. 6.1.

When discrepancies in the number of artifacts of interest are identified, the artifacts will be tagged, and their source file (database) and file system location will be reviewed to ensure the artifact is associated to a native iOS application as outlined below. Once an association to a native iOS application has been confirmed, the source database from the two extractions being compared will be reviewed to verify the data loss and to ensure that the forensic tools are not reporting false positives. Once removal is confirmed, the modified time of the appropriate database will be noted and used as part of the Timeline and iOS Analysis outlined below.

Any observed reductions in extraction size or number of artifacts will be an indication of data loss. In addition, subsequent reductions noted in the second and/or third comparisons will show that the longer an extraction is delayed, the greater the loss of data is.

Fig. 6.1 Comparison analysis outline

Table 6.1 iOS database locations

Artifacts of interest	Path
Electronic communications—calls	/private/var/wireless/Library/CallHistoryDB/Callhistory.storedata
Electronic communications—chats, instant messages (SMS, MMS)	/private/var/mobile/Library/SMS/sms.db
Media files—pictures, videos	/private/var/mobile/Media/PhotoData/Photos.sqlite
Logs—application usage log	/private/var/mobile/Library/CoreDuet/Knowledge/knowledgeC.db
Logs—log entries	/private/var/mobile/Library/CoreDuet/People/interactionC.db /private/var/wireless/Library/Databases/DataUsage.sqlite

Consistent reductions across multiple devices will demonstrate that a seized iPhone is a volatile container for its data as a result of the advanced forensic tools only supporting data extraction when the iPhone is kept powered on. These findings in turn, could be used to support investigations, demonstrating that the best possible outcome is achieved when an iPhone is extracted as soon as possible.

6.4.5.2   IOS Application/Database Analysis

Based on the artifacts of interest selected in Sect. 6.4.4, as part of the comparison analysis, certain applications and their respective databases are expected to be identified and reviewed. The iOS application databases including their file path locations within the iOS file system (based on the iOS versions of the iPhones used for this research) can be seen in Table 6.1.

It should be noted here, that only the data presented by the forensic analysis software will be reviewed. Additional data carving and the recovery of deleted artifacts outside of those recovered by the forensic software (including the recovery of entries from SQLite databases) will not form part of this research.

6.4.5.3   Timeline and iOS Analysis

When changes are made to a record in iOS, e.g. a recorded entry in a SQLite database is removed, the record's modified timestamp is updated to reflect the change. When viewing the modified time of a file in iOS, the modified time reflects the last (or most recent) time the file was modified. Armed with this information, if artifacts are identified as being lost when comparing the iPhone extractions, the modified time of the SQLite database where the artefact of interest was located can be cross-referenced in the timeline to determine what other activities were occurring on the device around the same time. This is possible because both forensic analysis tools have a timeline feature that has the ability to show all system and user events that occurred on a device in chronological order.

Additionally, since iOS version 10.X, Apple Inc. introduced the use of a unified logging system, which captures disk and memory log data from all system levels into one centralized location. These log files are located in the file system at:

- /private/var/db/diagnostics; and
- /private/var/db/uuidtext

Including these log files as part of the analysis and leveraging the timeline capabilities of the forensic analysis tools, it is expected that any loss of data can be further verified as well as potentially identifying the cause of the data loss.

6.5 Experiments and Findings

As mentioned in the previous section, five iPhone test devices were used in the experiments: SE, 5s, 7, X, 7+. At the start of the extraction phase, each iPhone was powered off. As a result, each device was powered on, and had their passcode entered into the lock screen to place them into an AFU or Hot state. The Airplane mode status was enabled, if required (ensuring Bluetooth and Wi-Fi connections were disabled), and the lock screens were subsequently enabled and verified. Each device.

As stated previously, the boot times of three of the devices are not current, as such, the date and time for each device is included as a reference in Table 6.2. Each iPhone had a seizure time selected and was connected to a charge cable, connected to a UPS in a secure room only accessible to the researcher. The timing of these actions are shown in Table 6.2 using Atlantic Daylight Time (ADT). All of the iPhones with the exception of the iPhone 7+ were showing a boot time in ADT. The iPhone 7+ was showing a time in Atlantic Standard Time (AST).

Table 6.2 Experimental devices

iPhone	Powered on time (ADT)	Seizure time (ADT)	Boot time (ADT)
SE	2021-06-15 12:50	2021-06-15 12:51	2016-09-21 19:45
5S	2021-06-15 14:30	2021-06-15 14:30	2021-06-15 14:30
7	2021-06-15 16:04	2021-06-15 16:04	2017-10-15 15:04
X	2021-06-15 17:33	2021-06-15 17:34	2021-06-15 17:33
7+	2021-06-19 05:51	2021-06-19 05:53	2018-01-01 14:16

The iOS application, database analysis and timeline iOS Analysis were carried out for 5 iPhone test devices, however only the results of iPhone 7 and X will be described in the following subsections. A comparison analysis of all testing devices will be presented at the end of this section.

6.5.1 iOS Application and Database Analysis

6.5.1.1 iPhone 7 Analysis

The only artifact of interest noted to have changed were the Images, shown in Table 6.3.

2 h versus 24 h

Images: One decrease in images was identified by the artifacts of interest comparison. In comparing the images from the iPhone 7's 2 h and 24 h extractions, one image file was identified in the 2 h extraction that was not found in the 24 h extraction. The image file was tagged and reviewed. The image file was associated to a source file of Cache.db-wal located in the iPhone 7's file system at /private/var/mobile/Library/Caches/sharedCaches/com.apple. iTunes-Store.NSURLCache/. The image was identified as an embedded file called 'Cache.db-wal_embedded_1.jpg'. With the iTunes Store falling outside the scope of the research paper, the embedded image was excluded from the research.

24 h versus 72 h

No changes occurred in the number of reported artifacts of interest reported.

72 h versus 7 days

Images: Comparing the artifacts of interest revealed a decrease of 48 images. The images from the iPhone 7's 72 h and 7 day extractions were compared, and the 48 images were identified in the 72 h extraction, tagged and reviewed. Of these 48 image files, 44 had an identified source path of /private/var/mobile/Library/Caches/com.apple.MobileSMS/Previews/Attachments/. As previously stated, com.apple.MobileSMS is the BundleID for the native iOS Messages application. These 44 images can be seen in Fig. 6.2.

With a native iOS application identified in the source path for 44 images, further analysis was conducted. The 44 lost images from this location all had

Table 6.3 iPhone 7 artefacts of interest comparison

Artefacts of interest	2 h	24 h	+/−	24 h	72 h	+/−	72 h	7 days	+/−
Images	42,826	42,825	−1	42,825	42,825	0	42,825	42,777	−48

#	Name	Size (Bytes)	Sub-Folders	MD5 Hash
1	20170731_212539-preview-l.ktx	101520	/2d/13/CAC20AC3-E9FD-4FAF-A867-ED2F0E21A273/	01b1eeb94dd1695c5160a90b684ea05f
2	2Uei0yTn-preview.ktx	322896	/9b/11/2CA52231-F734-4552-8F70-774A4D9AB73F/	af544fdbf0e939a2191146f1d5fac059
3	54cc296a542a130e3eebc2098d466499-preview.ktx	322896	/31/01/97D7A20B-93FC-403B-A1AD-48A528B2DB68/	158e520e18d0800a85ab96b1f254e2cc
8	Company-preview.ktx	242208	/d8/08/C51A9B03-BEBF-4312-BED2-6B6DAAFC6D56/	20f93a376b0a3519c20e2e506204e532
9	FB_IMG_1501636183546-preview.ktx	242208	/53/03/48B3ADAF-4716-46A7-ACFB-518417C27BD2/	1f7676979daa7847935be6b219d98110
10	GingerSnaps2-preview.ktx	242208	/bb/11/B14D97BC-7A12-42EE-8E9D-033511C82F59/	396e7dbe5db01404 5ecd33043e2a53ec
11	IMG_0248-preview.ktx	322896	/21/01/6640AA9F-C4CC-4B8A-9C78-025251D5766F/	ca4f4ca1bca50f0d91391f955dce902d
12	IMG_0293-preview.ktx	322896	/76/06/0CC34DCE-DE20-4109-9EA0-4B624C719E99/	ce3a2cb8371b2ec9d6aa354b128792c9
13	IMG_0294-preview.ktx	322896	/78/08/27E2AAA0-0237-43D1-AB74-01463BE52DBE/	665c3f276bf79602643c3ac4c8fbc785
14	IMG_0296-preview.ktx	322896	/2f/15/0FF1DDC9-23CA-4774-8909-EFEC7C627F74/	0faa197236fbfc4bcdf8e0f5c3638ef3
15	IMG_0298-preview.ktx	322896	/ef/15/B3A981A7-E752-4510-A794-6594FF7BB868/	9c97af95e7e4a5df5bfbba8f67b951d0
16	IMG_0299-preview.ktx	322896	/51/01/BF2979C8-4A35-49B7-8FD1-163BBADA62ED/	898127e8a6dcf117faadd682263d1c3a
17	IMG_0302-preview.ktx	322896	/42/02/381F3AD1-7DC6-49FA-A68D-06A103B8AC83/	d45366c67d0f7c65c41a447b914f6078
18	IMG_0310-preview.ktx	322896	/34/04/90F8094D-0631-493D-825C-595ACB2B499A/	86b43ddc92156a935e579a72375bebe7
19	IMG_0310-preview_1.ktx	322896	/b0/00/58965668-4C53-4100-B8AC-53361833A709/	86b43ddc92156a935e579a72375bebe7
20	IMG_0312-preview.ktx	181200	/40/00/210ED94E-201D-49E2-93EA-BCFCEF708A1C/	8f4f4af0e4a2c60e54994c4d086c7097
21	IMG_0312-preview_1.ktx	181200	/87/07/D072998C-1E52-4815-AD6B-0D61BED7EA60/	8f4f4af0e4a2c60e54994c4d086c7097
22	IMG_0313-preview.ktx	181200	/68/08/A990C377-0B02-4A54-96D7-78D036FB8782/	a9700d1cbbee3136b09e3639860327a5
23	IMG_0313-preview_1.ktx	181200	/7c/12/12A694C1-1B1B-4325-9D0C-21517D031A66/	a9700d1cbbee3136b09e3639860327a5
24	IMG_0314-preview.ktx	181200	/fd/13/502D5805-5C1D-4CC4-819B-8FCF26A9F6AC/	77b6ca03b0dcc6c0bca571030aeb5aef
25	IMG_0318-preview.ktx	322896	/e0/15/D1721B12-CFBD-44D5-B4F3-610295AFF0B0/	4e0577d8dacaa900a8a1a75dc1426300
26	IMG_0318-preview_1.ktx	322896	/3f/15/C2BEF5DA-B04A-4066-8D30-DCB31E75E8EE/	4e0577d8dacaa900a8a1a75dc1426300
27	IMG_0322-preview.ktx	181200	/bc/12/CE45112D-A1DD-431A-A53A-01903C710BCA/	942eabf4db1246c0cc0906e004635b4cd
28	IMG_0329-preview.ktx	322896	/4d/13/B39077DB-9A6A-419E-B0ED-D5FFABBFAF85/	c3d77e139beee2572a97c3f62099d422
29	IMG_0330-preview.ktx	322896	/1f/15/04E6FBAC-39F2-4DRF-9235-021E91D88FF4/	30a20fb57dceac9d4e3e1794a29445f4
30	IMG_0333-preview.ktx	322896	/02/02/566DE426-196F-4737-A42D-0897E286B636/	b567f34dfc2e54eb28c77e98c5fafa63
31	IMG_1704-preview.ktx	322896	/34/04/0C5BA154-D869-4848-BEA3-822BF10A0973/	208b83714033940f32c1bde3ff63a0f2
32	IMG_1744-preview.ktx	181200	/1c/12/CEB74AE1-FDFF-40C5-8583-1DCD716B9FC1/	30d6cbf5963ae52496b84ca150412e61
33	IMG_1764-preview.ktx	181200	/0f/15/41631BC7-29EC-4751-B6DF-D5912FEB72E6/	13d5b53b948f41e19d6be2822a371069
34	IMG_1773-preview.ktx	181200	/dc/12/C653E662-33CA-4D50-8F58-E40CAE3C8A85/	c24dae47d08e3d8868ec9d24980c7d9b
35	Labels2-preview.ktx	242208	/f1/01/19564010-3D88-4742-B7CF-F83EE3D90055/	199a6de4539730abd684c93d29bc10c3
36	Resized_20170218_160005-preview.ktx	322896	/db/11/3FBF6AEE-B733-4112-B52F-D76262F3AE98/	d8630ebb294a7fa6c7e59aa61afcce3f
37	Resized_20170405_212421-preview.ktx	322896	/a6/06/2F2621-8C04-4F87-9C3A-BBE6FB3C89EE/	54b03fdae8a1a77d6d4051e0e2865ec8
38	Resized_20170725_153714-preview.ktx	322896	/24/04/32258725-CDED-44B9-B908-C10784577D33/	96d047196bf0257a77dc5bbc20b3fd6b
39	Resized_20170728_092101-preview.ktx	135936	/de/14/91BD9059-61DB-4D5B-A3D4-63E1F80B2600/	b18aad752836af7c535b55943ebc2d0
40	Resized_20170728_092447-preview.ktx	135936	/2e/14/8D35AE17-2CCA-4401-9CA5-EBDCA56F7BA6/	2db6eb647e56f4b6d174ad8f99beb3c2
41	Resized_20170728_175905-preview.ktx	135936	/8b/11/C0019EC0-5695-4389-97B8-38E3D0863B90/	b6753c5317e8a73f183d331cff42f368
42	Resized_20170728_175913-preview.ktx	135936	/55/05/C8E45965-CF23-441C-9FF7-20D4D251B369/	fc5599da00c7c8bc2226a5ed0b839d19
43	Resized_20170728_181309-preview.ktx	135936	/97/07/1986EFC7-85A7-4D07-85FB-BDF32B2C5732/	ad1df30a41a79fcb00cf8565294c496f
44	Resized_20170730_143115-preview.ktx	135936	/42/02/32F749FD-E261-4A01-9D8E-459A42FEA9C2/	ca77bdc34dc3b7d356735c0c06a518d9
45	Resized_20170730_143301-preview.ktx	191040	/bc/12/4DB276F7-1CB2-4566-9351-0B683A0BB3CA/	f9b1bac4844f48e05ef9dd1271d27e4d
46	Resized_IMG_20170728_103009133-preview.ktx	135936	/51/01/E14EEE6F-5F61-46B6-8434-BCA746FC62DA/	0c999391cee60c430a112289cfd77ee8
47	Screenshot_20170720-085049-preview.ktx	322896	/34/04/B24F019B-3E93-4654-B20F-B656CE394C03/	e31a38b8ddf389e2f6f1e1bb82b3a851
48	Untitled-26-preview.ktx	246144	/a6/06/8DAC44CD-F30A-4717-AEBB-E4B9154775DC/	d25fe91b1a0aa39fc9784f9bdd1829fc

Source Path: /private/var/mobile/Library/Caches/com.apple.MobileSMS/Previews/Attachments/

Fig. 6.2 iPhone 7—44 images lost from 72 h extraction

the same file extension of '.ktx' which is a known file extension for iOS Snapshots. iOS Snapshots are tracked in the applicationState.db database located at /private/var/mobile/Library/FrontBoard/. Reviewing this database from the iPhone 7's 72 h extraction did not reveal any entries that correlated to the 44 image files. Based on this finding, it appears that the purpose of these '.ktx' images is like a thumbnail image, in that each '.ktx' preview is a smaller version of an SMS image attachment.

In reviewing the subfolder location for each of the 44 lost images, removal of each of the 44 images was confirmed, and in doing so, a pattern was established. Each sub-subfolder that contained one of the 44 '.ktx' images had been removed and each parent subfolder had the same modified time of 2017-10-22 03:00. A comparison of the file path /private/var/mobile/Library/Caches/com.apple.MobileSMS/Previews/Attachments/2d/13/ from the 72 h and 7 day extractions

Name	Size	Modified	Name	Size	Modified
13	101 520	2017-08-01 14:14	13	0	2017-10-22 03:00

Fig. 6.3 iPhone 7 iOS snapshot "/Previews/Attachments/2d/13/" 72 h versus 7 days

where the CAC20AC3-E9FD-4FAF-A867-ED2F0E21A273/20170731_212539-preview-l.ktx file was located in the 72 h extraction, is shown in Fig. 6.3.

In comparing the overall size difference of the Attachments folder from the iPhone 7's 72 h and 7 day extractions, although there was no change in modified time, the Attachments folders size shows a decrease of 11,121,749 B, displayed in Fig. 6.4.

The other four of the 48 images had an associated source file of the Cache.db-wal write ahead log located in the iPhone 7's file system at /private/var/mobile/Library/Caches/com.apple.parsecd/ Cache.db-wal. A list of the four images is presented in Fig. 6.5.

The forensic analysis software identified each of the four photos as embedded '.jpg' files, and of the four, three had the word 'partial' in its name, indicating that they were likely only partial images (possibly meaning they are partially overwritten). A comparison of the Cacbe.db-wal from the 72 h and 7 day extractions is displayed in Fig. 6.6.

Name	Size	Modified	Name	Size	Modified
Attachments	35 980 042	2017-08-29 07:13	Attachments	24 786 293	2017-08-29 07:13

Fig. 6.4 iPhone 7 attachments folder comparison 72 h versus 7 days

#	Name	Size (Bytes)	Path	MD5 Hash
4	Cache.db-wal_embedded_1_partial.jpg	2552	/private/var/mobile/Library/Caches/com.apple.parsecd/Cache.db-wal/	1607e7955a2696a0a9b32d325d81e355
5	Cache.db-wal_embedded_2_partial.jpg	976	/private/var/mobile/Library/Caches/com.apple.parsecd/Cache.db-wal/	d6e899268b5ecdb4ec96a15be40fed35
6	Cache.db-wal_embedded_3.jpg	3829	/private/var/mobile/Library/Caches/com.apple.parsecd/Cache.db-wal/	8a66f9ead48c46fba9393e12287f642c
7	Cache.db-wal_embedded_4_partial.jpg	21983	/private/var/mobile/Library/Caches/com.apple.parsecd/Cache.db-wal/	8bd1d874b84aba7168618487a6bcb120

Fig. 6.5 iPhone 7—4 lost images from 72 h extraction

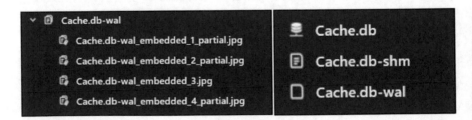

Fig. 6.6 iPhone 7 Cacbe.db-wal comparison 72 h versus 7 days showing 4 embedded images

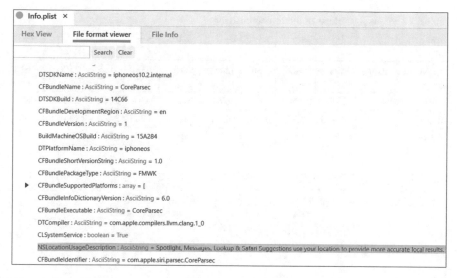

Fig. 6.7 iPhone 7 Cacbe.db-wal comparison 72 h versus 7 days showing 4 embedded images

According to the Info.plist located in /System/Library/PrivateFrameworks /CoreParsec.framework/Versions/Current/Resources/, the com.apple.parsecd daemon is associated the phone's location to assist with lookup, spotlight and safari suggestions, shown in Fig. 6.7.

Since these four embedded image files were not associated to one of the native iOS application included in the scope of this research paper, they were excluded from the research findings.

Briefly, the iPhone 7 only saw a decrease of 49 images between the four timed extractions. Of these 49 images, five were excluded as they were associated to applications or databases that fell outside the scope of this research paper. The remaining 44 images were associated to the native iOS Messages application, and were noted to all be '.ktx' files. The removal of each of the 44 images was confirmed through the analysis conducted.

6.5.1.2   iPhone X Analysis

The changes noted through the artifacts of interest comparison were the Application Usage Log, Log Entries, Images and Videos, shown in Table 6.4.

With decreases identified in the Application Usage Log and Log entries, the knowledgeC.db, interactionC.db and DataUsage.sqlite databases were analysed to determine if the auto-vacuum mode was enabled. The Z_METADATA tables from all three databases were reviewed to look at the stored binary '.plist' file contained in the tables. The binary '.plist' file for the knowledgeC.db confirmed the database

Table 6.4 iPhone X artefacts of interest comparison

Artefacts of interest	2 h	24 h	+/−	24 h	72 h	+/−	72 h	7 days	+/−
Applications usage log	3608	3245	−363	3245	3115	−130	3115	2500	−615
Log entries	2719	2678	−41	2678	2642	−36	2642	2555	−87
Images	34,848	34,848	0	34,848	34,561	−287	34,561	34,495	−66
Videos	613	613	0	613	556	−57	556	546	−10

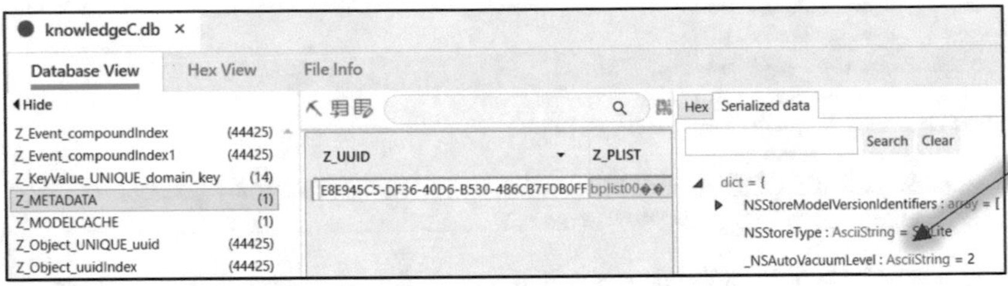

Fig. 6.8 iPhone X knowledgeC.db auto-vacuum mode

was running in INCREMENTAL auto-vacuum mode, identified by the number '2' next to the _NSAutoVacuumLevel, shown in Fig. 6.8.

The binary '.plist' file from the interactionC.db database confirmed that it too was running in INCREMENTAL auto-vacuum mode, identified by the number '2' next to the _NSAutoVacuumLevel, depicted in Fig. 6.9.

The binary '.plist' file from the DataUsate.sqlite database confirmed that it was also running in INCREMENTAL auto-vacuum mode, identified by the number '2' next to the _NSAutoVacuumLevel, depicted in Fig. 6.10.

2 h versus 24 h

Application Usage Log: A decrease in 363 entries was reported in the artefacts of interest comparison. The Application Usage Log artifacts were compared from the iPhone X's 2 h and 24 h extractions, and the 363 entries were identified in the 2 h extraction, tagged and reviewed. Due to the high number of entries, we do not list Application Usage Log entries.

Of the 363 entries, 179 had an associated source file of the knowledgeC.db (ZOBJECT, ZSTRUCTUREMETADATA tables) database while the other 184 had an associated source file of the knowledgeC.db (ZOBECT table) database. Comparing the knowledgeC.db database files from the iPhone X's 2 h and 24 h extractions, a new modified time of 2021-06-16 17:46 as well as a decrease in size of 1,036,288 B was noted for the knowledgeC.db database in the 24 h extraction, seen in Fig. 6.11.

The reduction in the size of the knowledgeC.db supports the decrease in 363 log entries, and the reduction in size of the database is supported by the knowledgeC.db database running in incremental auto-vacuum mode.

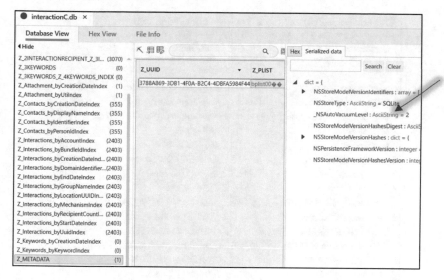

Fig. 6.9 iPhone X interactionC.db auto-vacuum mode

Fig. 6.10 iPhone X DataUsage.sqlite incremental auto-vacuum mode

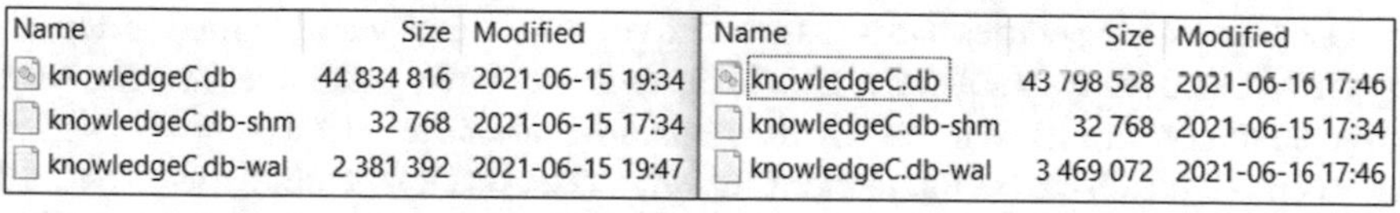

Name	Size	Modified	Name	Size	Modified
knowledgeC.db	44 834 816	2021-06-15 19:34	knowledgeC.db	43 798 528	2021-06-16 17:46
knowledgeC.db-shm	32 768	2021-06-15 17:34	knowledgeC.db-shm	32 768	2021-06-15 17:34
knowledgeC.db-wal	2 381 392	2021-06-15 19:47	knowledgeC.db-wal	3 469 072	2021-06-16 17:46

Fig. 6.11 iPhone X knowledgeC.db database files comparison 2 h versus 24 h

In analyzing the ZOBJECTS table in the knowledgeC.db database from the 2 h and 24 h extractions, a reduction in the number of entries from 44,425 to 40,731 was noted, shown in Fig. 6.12.

Comparing the entries in each of the ZOBJECTS tables, the removal of the 363 Application Usage Log entries from the 2 h extraction was confirmed. The entries were confirmed to have been removed between entry numbers 14,040 and 18,204.

Log Entries: The comparison of artifacts of interest identified a decrease in 41 log entries between the iPhone X's 2 h and 24 h extractions. The reported artifacts

knowledgeC.db × ○ kno		knowledgeC.db × ○ kno
Database View	Hex View	**Database View**	Hex View
◀ Hide		◀ Hide

Z_Event_compoundIndex	(44425)	Z_Event_compoundIndex	(40731)
Z_Event_compoundIndex1	(44425)	Z_Event_compoundIndex1	(40731)
Z_KeyValue_UNIQUE_domain_key	(14)	Z_KeyValue_UNIQUE_domain_key	(14)
Z_METADATA	(1)	Z_METADATA	(1)
Z_MODELCACHE	(1)	Z_MODELCACHE	(1)
Z_Object_UNIQUE_uuid	(44425)	Z_Object_UNIQUE_uuid	(40731)
Z_Object_uuidIndex	(44425)	Z_Object_uuidIndex	(40731)
Z_PRIMARYKEY	(17)	Z_PRIMARYKEY	(17)
Z_StructuredMetadata_UNIQUE_...	(7194)	Z_StructuredMetadata_UNIQUE_...	(6774)
Z_SyncPeer_UNIQUE_uuid	(1)	Z_SyncPeer_UNIQUE_uuid	(1)
ZADDITIONCHANGESET	(0)	ZADDITIONCHANGESET	(0)
ZCONTEXTUALCHANGEREGISTRATI...	(109)	ZCONTEXTUALCHANGEREGISTRATI...	(109)
ZCONTEXTUALKEYPATH	(0)	ZCONTEXTUALKEYPATH	(0)
ZCUSTOMMETADATA	(2339)	ZCUSTOMMETADATA	(2205)
ZCUSTOMMETADATA_ZOBJECT_I...	(2339)	ZCUSTOMMETADATA_ZOBJECT_I...	(2205)
ZDELETIONCHANGESET	(0)	ZDELETIONCHANGESET	(0)
ZHISTOGRAM	(7)	ZHISTOGRAM	(7)
ZHISTOGRAMVALUE	(338)	ZHISTOGRAMVALUE	(355)
ZHISTOGRAMVALUE_ZHISTOGRAM...	(338)	ZHISTOGRAMVALUE_ZHISTOGRAM...	(355)
ZKEYVALUE	(14)	ZKEYVALUE	(14)
ZOBJECT	(44425)	ZOBJECT	(40731)

Fig. 6.12 iPhone X ZOBJECT table comparison 2 h versus 24 h

were compared and the 41 entries were identified in the 2 h extraction, tagged and reviewed. A listing of the 41 entries can be seen in Fig. 6.13.

One of the 41 log entries had no associated source file and was identified as deleted by the forensic analysis software. Due to the entry not having a source file, it was not possible to confirm its removal. Of the remaining 40 entries, 10 were associated to the interactionC.db-wal or interactionC.db (ZINTERACTIONS table) files. These 10 entries were associated to the com.apple.InCallService application. The remaining 30 entries had an associated source file of the DataUsage.sqlite-wal (ZLIVEUSAGE, ZPROCESS tables) database. When comparing the interactionC.db database from the iPhone X's 2 h and 24 h extractions, there was no new modified time or change in database size noted, seen in Fig. 6.14.

However, when comparing the ZINTERACTIONS tables in the interactionC.db databases from the 2 h and 24 h extractions, a decrease in entries from 2403 to 2393 was noted, shown in Fig. 6.15. This decrease in 10 entries does support the identified decrease from the artifacts of interest comparison.

In reviewing the contents of both ZINTERACTIONS tables, the reduction in 10 log entries was confirmed. Using a filter of 2021-05-19 in the ZENDATE col-umn and filter of com.apple.InCallService in the ZBUNDLEID column in both tables,

#	Application	Body	Start Time (UTC-3)	End time (UTC-3)	Source file information
1	com.apple.datausage.appleid	Wifi In:0 Wifi Out:0Wan In:24686 Wan Out:5801	2021-06-02 18:19		DataUsage.sqlite-wal : 0x376D9D (Table: ZLIVEUSAGE, ZPROCESS)
2	com.apple.datausage.siri	Wifi In:0 Wifi Out:0Wan In:31620 Wan Out:7308	2021-06-02 18:19		DataUsage.sqlite-wal : 0x376E3C (Table: ZLIVEUSAGE, ZPROCESS)
3	com.apple.datausage.security	Wifi In:0 Wifi Out:0Wan In:15703 Wan Out:4423	2021-06-02 18:18		DataUsage.sqlite-wal : 0x3768A6 (Table: ZLIVEUSAGE, ZPROCESS)
4	Safari	Wifi In:0 Wifi Out:0Wan In:14402997 Wan Out:847826	2021-05-31 21:16		DataUsage.sqlite-wal : 0x377D81 (Table: ZLIVEUSAGE, ZPROCESS)
5	Safari	Wifi In:0 Wifi Out:0Wan In:9705 Wan Out:17548	2021-05-31 21:16		DataUsage.sqlite-wal : 0x377DA3 (Table: ZLIVEUSAGE, ZPROCESS)
6	com.apple.datausage.siri	Wifi In:0Wifi Out:0Wan In:560631Wan Out:1504156	2021-05-29 20:34		DataUsage.sqlite-wal : 0x377BFB (Table: ZLIVEUSAGE, ZPROCESS)
7	com.apple.datausage.location	Wifi In:0Wifi Out:0Wan In:10106982Wan Out:688865	2021-05-29 19:44		DataUsage.sqlite-wal : 0x377479 (Table: ZLIVEUSAGE, ZPROCESS)
8	com.apple.AppStore	Wifi In:0Wifi Out:0Wan In:72154Wan Out:40987	2021-05-29 19:44		DataUsage.sqlite-wal : 0x377956 (Table: ZLIVEUSAGE, ZPROCESS)
9	News	Wifi In:0Wifi Out:0Wan In:82411Wan Out:27246	2021-05-29 19:44		DataUsage.sqlite-wal : 0x377977 (Table: ZLIVEUSAGE, ZPROCESS)
10	com.apple.datausage.iad	Wifi In:0Wifi Out:0Wan In:114981Wan Out:42149	2021-05-29 19:44		DataUsage.sqlite-wal : 0x377D60 (Table: ZLIVEUSAGE, ZPROCESS)
11	com.apple.datausage.findmyiphone	Wifi In:0Wifi Out:0Wan In:168457Wan Out:168791	2021-05-29 19:44		DataUsage.sqlite-wal : 0x377997 (Table: ZLIVEUSAGE, ZPROCESS)
12	com.apple.AppStore	Wifi In:0Wifi Out:0Wan In:5124Wan Out:6351	2021-05-29 19:44		DataUsage.sqlite-wal : 0x377F9F9 (Table: ZLIVEUSAGE, ZPROCESS)
13	com.apple.datausage.general	Wifi In:0Wifi Out:0Wan In:218484Wan Out:88603	2021-05-29 19:44		DataUsage.sqlite-wal : 0x377A18 (Table: ZLIVEUSAGE, ZPROCESS)
14	com.apple.datausage.itunesmedia	Wifi In:0Wifi Out:0Wan In:110946Wan Out:43536	2021-05-29 19:44		DataUsage.sqlite-wal : 0x377B8 (Table: ZLIVEUSAGE, ZPROCESS)
15	News	Wifi In:0Wifi Out:0Wan In:398129Wan Out:187982	2021-05-29 17:19		DataUsage.sqlite-wal : 0x377B7C (Table: ZLIVEUSAGE, ZPROCESS)
16	com.apple.datausage.appleid	Wifi In:0Wifi Out:0Wan In:423694Wan Out:132948	2021-05-29 17:19		DataUsage.sqlite-wal : 0x377935 (Table: ZLIVEUSAGE, ZPROCESS)
17	com.apple.datausage.docsandsync	Wifi In:0Wifi Out:0Wan In:171960Wan Out:40963	2021-05-29 17:19		DataUsage.sqlite-wal : 0x37749B (Table: ZLIVEUSAGE, ZPROCESS)
18	Weather	Wifi In:0Wifi Out:0Wan In:173240Wan Out:63492	2021-05-29 17:19		DataUsage.sqlite-wal : 0x3778D6 (Table: ZLIVEUSAGE, ZPROCESS)
19	Weather	Wifi In:0Wifi Out:0Wan In:26018Wan Out:10819	2021-05-29 17:19		DataUsage.sqlite-wal : 0x3773F7 (Table: ZLIVEUSAGE, ZPROCESS)
20	Maps	Wifi In:0Wifi Out:0Wan In:128854Wan Out:32901	2021-05-29 17:19		DataUsage.sqlite-wal : 0x377458 (Table: ZLIVEUSAGE, ZPROCESS)
21	Weather	Wifi In:0Wifi Out:0Wan In:22180Wan Out:5864	2021-05-29 17:19		DataUsage.sqlite-wal : 0x3778F7 (Table: ZLIVEUSAGE, ZPROCESS)
22	com.apple.datausage.itunesmedia	Wifi In:0Wifi Out:0Wan In:227663Wan Out:29995	2021-05-29 17:19		DataUsage.sqlite-wal : 0x3782DB (Table: ZLIVEUSAGE, ZPROCESS)
23	com.apple.datausage.itunesmedia	Wifi In:0Wifi Out:0Wan In:1033513Wan Out:176127	2021-05-29 17:19		DataUsage.sqlite-wal : 0x377A59 (Table: ZLIVEUSAGE, ZPROCESS)
24	News	Wifi In:0Wifi Out:0Wan In:2306012Wan Out:209519	2021-05-29 17:19		DataUsage.sqlite-wal : 0x377B3C (Table: ZLIVEUSAGE, ZPROCESS)
25	Maps	Wifi In:0Wifi Out:0Wan In:6801Wan Out:6792	2021-05-29 17:19		DataUsage.sqlite-wal : 0x377A7A (Table: ZLIVEUSAGE, ZPROCESS)
26	com.apple.datausage.softwareupdate	Wifi In:0Wifi Out:0Wan In:98767Wan Out:35099	2021-05-29 17:19		DataUsage.sqlite-wal : 0x377DE3 (Table: ZLIVEUSAGE, ZPROCESS)
27	News	Wifi In:0Wifi Out:0Wan In:19824Wan Out:17309	2021-05-29 17:19		DataUsage.sqlite-wal : 0x377B5D (Table: ZLIVEUSAGE, ZPROCESS)
28	com.apple.datausage.dns	Wifi In:0Wifi Out:0Wan In:225626Wan Out:221979	2021-05-29 17:19		DataUsage.sqlite-wal : 0x377E24 (Table: ZLIVEUSAGE, ZPROCESS)
29	FaceTime	Wifi In:0Wifi Out:0Wan In:5588Wan Out:5588	2021-05-29 13:55		DataUsage.sqlite-wal : 0x377ABC (Table: ZLIVEUSAGE, ZPROCESS)
30	com.apple.datausage.applepushservice	Wifi In:0Wifi Out:0Wan In:2106483Wan Out:3592809	2021-05-29 5:23		DataUsage.sqlite-wal : 0x377BDA (Table: ZLIVEUSAGE, ZPROCESS)
31	com.apple.InCallService	incoming call	2021-05-19 12:08	2021-05-19 12:08	interactionC.db : 0xA7D10 (Table: ZINTERACTIONS)
32	com.apple.InCallService	incoming call	2021-05-19 11:45	2021-05-19 11:55	interactionC.db : 0xA7FA8 (Table: ZINTERACTIONS)
33	com.apple.InCallService	outgoing call	2021-05-19 11:00	2021-05-19 11:00	interactionC.db : 0x324266 (Table: ZINTERACTIONS)
34	com.apple.InCallService	incoming call	2021-05-19 11:00	2021-05-19 11:00	interactionC.db : 0x3242F6 (Table: ZINTERACTIONS)
35	com.apple.InCallService	outgoing call	2021-05-19 10:20	2021-05-19 10:21	interactionC.db : 0x32437A (Table: ZINTERACTIONS)
36	com.apple.InCallService	outgoing call	2021-05-19 9:18	2021-05-19 9:18	interactionC.db : 0x3247B3 (Table: ZINTERACTIONS)
37	com.apple.InCallService	outgoing call	2021-05-19 9:16	2021-05-19 9:16	interactionC.db : 0x32A41E (Table: ZINTERACTIONS)
38	com.apple.InCallService	outgoing call	2021-05-19 9:15	2021-05-19 9:15	interactionC.db : 0x324F93 (Table: ZINTERACTIONS)
39	com.apple.InCallService	outgoing call	2021-05-19 9:15	2021-05-19 9:15	interactionC.db : 0x324024 (Table: ZINTERACTIONS)
40	com.apple.InCallService	outgoing call	2021-05-19 9:15	2021-05-19 9:15	interactionC.db : 0x3240B5 (Table: ZINTERACTIONS)
41		call			

Fig. 6.13 iPhone X lost log entries 2 h extraction

Name	Size	Modified	Name	Size	Modified
interactionC.db	1 572 864	2021-06-11 16:48	interactionC.db	1 572 864	2021-06-11 16:48
interactionC.db-shm	32 768	2021-06-15 18:15	interactionC.db-shm	32 768	2021-06-16 17:39
interactionC.db-wal	3 328 992	2021-06-15 18:15	interactionC.db-wal	3 427 872	2021-06-16 17:39

Fig. 6.14 iPhone X interactionC.db database files comparison 2 h versus 24 h

the removed entries were identified as entry numbers 1033–1035, 1037, 1041, and 1048–1052, which can be seen in Fig. 6.16.

With the removal of the 10 log entries from the interactionC.db confirmed, analysis was conducted on the remaining 30 log entries associated to the DataUsage.sqlite-wal (ZLIVEUSAGE, ZPROCESS tables) write ahead lot. When comparing the DataUsage.sqlite database files from the iPhone X's 2 h and 24 h extractions, a new modified time of 2021-06-16 17:34 in the 24 h extraction for DataUsage.sqlite and DataUsage.sqlite-wal files was noted along with a reduction in size of 3,411,360 B for the DataUsage.sqlite-wal write ahead log. These changes can be seen in Fig. 6.17.

In reviewing the ZLIVEUSAGE table in the DataUsage.sqlite database from both extractions, a reduction in table entries was noted from 289 to 258, shown in Fig. 6.18.

Comparing the ZLIVEUSAGE entries from both tables, using a date filter of 2021-05-29 to 2021-06-02 in the ZTIMESTAMP column, removal of the 30 entries was able to be confirmed as entry numbers 440–442 and 444–470 (Fig. 6.19).

Although the removal of these entries was confirmed from the DataUsage.sqlite, the original identified entries were associated to the DataUsage.sqlite-wal write ahead log. Therefore, to ensure the entries were actually removed, the hex viewer in the

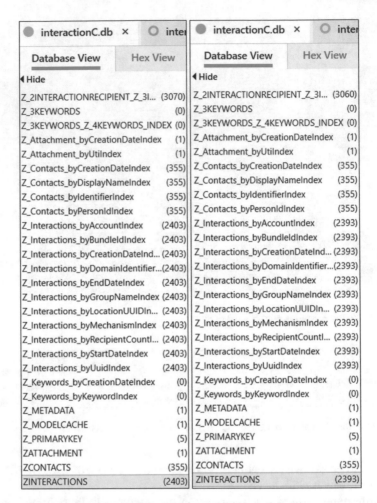

Fig. 6.15 iPhone X ZINTERACTIONS table comparison 2 h versus 24 h

Z_PK	ZENDDATE	ZSTARTDATE	ZBUNDLEID	Z_PK	ZENDDATE	ZSTARTDATE	ZBUNDLEID
1052	2021-05-19 3:08:22 PM	2021-05-19 3:08:22 PM	com.apple.InCallService				
1051	2021-05-19 2:55:33 PM	2021-05-19 2:45:26 PM	com.apple.InCallService				
1050	2021-05-19 2:00:15 PM	2021-05-19 2:00:15 PM	com.apple.InCallService				
1049	2021-05-19 2:00:08 PM	2021-05-19 2:00:08 PM	com.apple.InCallService				
1048	2021-05-19 1:21:15 PM	2021-05-19 1:20:37 PM	com.apple.InCallService				
1041	2021-05-19 12:18:24 PM	2021-05-19 12:18:24 PM	com.apple.InCallService				
1037	2021-05-19 12:16:36 PM	2021-05-19 12:16:36 PM	com.apple.InCallService				
1035	2021-05-19 12:15:45 PM	2021-05-19 12:15:45 PM	com.apple.InCallService				
1034	2021-05-19 12:15:30 PM	2021-05-19 12:15:30 PM	com.apple.InCallService				
1033	2021-05-19 12:15:25 PM	2021-05-19 12:15:25 PM	com.apple.InCallService				

Fig. 6.16 iPhone X ZINTERACTIONS table comparison identifying removed log entries

Name	Size	Modified	Name	Size	Modified
DataUsage.sqlite	126 976	2021-06-11 19:31	DataUsage.sqlite	126 976	2021-06-16 17:34
DataUsage.sqlite-shm	32 768	2021-06-15 17:33	DataUsage.sqlite-shm	32 768	2021-06-15 17:33
DataUsage.sqlite-wal	3 637 992	2021-06-15 17:34	DataUsage.sqlite-wal	226 632	2021-06-16 17:34

Fig. 6.17 iPhone X DataUsage.sqlite database files comparison 2 h versus 24 h

DataUsage.sqlite × ○ Da	DataUsage.sqlite × ○ Da
Database View	Hex View	**Database View**	Hex View
◀ Hide	◀ Hide

sqlite_master	(20)	sqlite_master	(20)
Z_METADATA	(1)	Z_METADATA	(1)
Z_MODELCACHE	(1)	Z_MODELCACHE	(1)
Z_PRIMARYKEY	(9)	Z_PRIMARYKEY	(9)
ZCHECKUPEVENT	(0)	ZCHECKUPEVENT	(0)
ZDEMOLIVEUSAGE	(0)	ZDEMOLIVEUSAGE	(0)
ZEVENT	(0)	ZEVENT	(0)
ZEVENT_ZHAPPENEDONNET_INDEX	(0)	ZEVENT_ZHAPPENEDONNET_INDEX	(0)
ZEVENT_ZHASPEER_INDEX	(0)	ZEVENT_ZHASPEER_INDEX	(0)
ZEVENT_ZHASSCENE_INDEX	(0)	ZEVENT_ZHASSCENE_INDEX	(0)
ZEVENTSCENE	(0)	ZEVENTSCENE	(0)
ZEVENTSCENE_ZWITHEVENT_INDEX	(0)	ZEVENTSCENE_ZWITHEVENT_INDEX	(0)
ZLIVEUSAGE	(289)	ZLIVEUSAGE	(258)

Fig. 6.18 iPhone X comparison of ZLIVEUSAGE tables in DataUsage.sqlite database

forensic analysis software was used to review the DataUsage.sqlite-wal files. The ending offset for the DataUsage.sqlite-wal from the 2 h extraction was noted to be $0 \times 37{,}548$, and the ending offset for the DataUsage.sqlite-wal file from the 24 h extraction was noted to be $0 \times 34782E8$, seen in Fig. 6.20.

Based on this finding, the starting offsets for the 30 log entries associated to the DataUsage.sqlite-wal file were reviewed again. The starting offsets ranged from $0 \times 3768A6$ to $0 \times 377DA3$, which occur after the end of file offset of $0 \times 37{,}548$ for the DataUsage.sqlite-wal from the 24 h extraction. Coupling this information with the reduction in size of the write ahead log seen in Fig. 6.19, the removal of the 30 log entries could be confirmed.

24 h versus 72 h

Application Usage Log: A decrease in 130 entries was identified by the comparison of artifacts of interest. The reported artifacts from the iPhone X's 24 h and 72 h extractions were compared, and the 130 entries were identified in the 2 h extraction, tagged and reviewed. Of the 130 entries, 124 had an associated source file of the knowledgeC.db (ZOBJECT table) database, 5 had an associated source file of the knowledgeC.db (ZOBJECT, ZSTRUCTUREMETADATA tables) database one

Z_PK	ZHASPROCESS	ZTIMESTAMP	ZWWANIN	ZWWANOUT
398	109	2021-05-31 22:28	11221	6569
426	110	2021-06-02 1:17	596	342
427	111	2021-06-02 1:17	10298	5232
440	74	2021-06-02 14:55	1866	5056
441	75	2021-06-02 14:55	6808301	180827
442	43	2021-06-02 18:38	41055	78172
443	36	2021-06-02 21:18	903	257
444	91	2021-06-02 21:18	15703	4423
445	50	2021-06-02 21:18	88311	18695
446	36	2021-06-02 21:18	141606	96680
447	90	2021-06-02 21:18	661	870
448	96	2021-06-02 21:18	21775	7272
449	99	2021-06-02 21:18	41024	11331
450	51	2021-06-02 21:18	1700020	141456
451	94	2021-06-02 21:18	23715	19375
452	105	2021-06-02 21:18	16048	4677
453	98	2021-06-02 21:18	11788	5405
454	88	2021-06-02 21:18	49773	17133
455	93	2021-06-02 21:18	15628	4070
456	97	2021-06-02 21:18	13728	4425
457	48	2021-06-02 21:19	33809	5297
458	54	2021-06-02 21:19	2550	2608
459	55	2021-06-02 21:19	768803	40743
460	77	2021-06-02 21:19	634	430
461	101	2021-06-02 21:19	3618	2863
462	78	2021-06-02 21:19	7815	3671
463	40	2021-06-02 21:19	36091	10046
464	103	2021-06-02 21:19	44082	13033
465	107	2021-06-02 21:19	24686	5801
466	83	2021-06-02 21:19	31620	7308
467	81	2021-06-02 21:19	160655	31122
468	100	2021-06-02 21:19	44739	7609
469	108	2021-06-02 21:19	15853	6340
783	39	2021-05-30 6:00	81482	32407
796	73	2021-05-30 6:00	69813	31341
798	75	2021-06-01 0:16	14402997	847826
801	80	2021-05-29 21:48	11532703	5228070
817	104	2021-05-30 15:49	1678217	165760

Z_PK	ZHASPROC	ZTIMESTAMP	ZWWANIN	ZWWANOUT
398	109	2021-05-31 10:28:54 PM	11221	6569
426	110	2021-06-02 1:17:31 AM	596	342
427	111	2021-06-02 1:17:33 AM	10298	5232
443	36	2021-06-02 9:18:08 PM	903	257
783	39	2021-05-30 6:00:35 AM	81482	32407
796	73	2021-05-30 6:00:35 AM	69813	31341
801	80	2021-05-29 9:48:28 PM	11532703	5228070
817	104	2021-05-30 3:49:14 PM	1678217	165760

Fig. 6.19 iPhone X ZLIVEUSAGE table comparison showing lost entries 2 h versus 24 h

```
00037530   00 00 00 00 00 00 00 00 00 00 00 00 00 00 00 00    ................
00037540   00 00 00 00 00 00 00 00                            ........

003782D0   08 09 01 00 07 08 08 03 02 05 64 41 C3 31 6C B4    ..........dA.1l.
003782E0   51 8F 1D 03 79 4F 75 2B                            Q...yOu+
```

Fig. 6.20 iPhone X end of file offset 0 × 34783E8 comparison for the DataUsage.sqlite-wal 2 h versus 24 h

had an associated source file of both the knowledgeC.db (ZOBJECT table) and the knowledgeC.db-wal (ZSTRUCTUREDMETADATA table) files. Note that not all 130 Application Usage Log entries are listed.

Comparing the knowledgeC.db database files from the iPhone X's 24 h and 72 h extractions, a new modified time of 2021-06-18 17:37 and a reduction in size of 2,678,784 B was noted in the 72 h extraction for the knowledgeC.db database. In addition, a new modified time of 2021-06-18 17:46 and increase in size of 646,840 B was noted in the 72 h extraction for the knowledgeC.db-wal write ahead log. These changes can be seen in Fig. 6.21.

Reviewing the knowledgeC.db database ZOBJECT tables from both extractions, a reduction in entries from 40,731 to 39,364 was observed, seen in Fig. 6.22.

Name	Size	Modified	Name	Size	Modified
knowledgeC.db	43 798 528	2021-06-16 17:46	knowledgeC.db	41 119 744	2021-06-18 17:37
knowledgeC.db-shm	32 768	2021-06-15 17:34	knowledgeC.db-shm	32 768	2021-06-15 17:34
knowledgeC.db-wal	3 469 072	2021-06-16 17:46	knowledgeC.db-wal	4 115 912	2021-06-18 17:46

Fig. 6.21 iPhone X knowledgeC.db database files comparison 24 h versus 72 h

knowledgeC.db ×		knowledgeC.db ×	
Database View	Hex View	**Database View**	Hex View
◀ Hide		◀ Hide	
Z_Event_compoundIndex	(40731)	Z_Event_compoundIndex	(39364)
Z_Event_compoundIndex1	(40731)	Z_Event_compoundIndex1	(39364)
Z_KeyValue_UNIQUE_domain_key	(14)	Z_KeyValue_UNIQUE_domain_key	(14)
Z_METADATA	(1)	Z_METADATA	(1)
Z_MODELCACHE	(1)	Z_MODELCACHE	(1)
Z_Object_UNIQUE_uuid	(40731)	Z_Object_UNIQUE_uuid	(39364)
Z_Object_uuidIndex	(40731)	Z_Object_uuidIndex	(39364)
Z_PRIMARYKEY	(17)	Z_PRIMARYKEY	(17)
Z_StructuredMetadata_UNIQUE_...	(6774)	Z_StructuredMetadata_UNIQUE_...	(6457)
Z_SyncPeer_UNIQUE_uuid	(1)	Z_SyncPeer_UNIQUE_uuid	(1)
ZADDITIONCHANGESET	(0)	ZADDITIONCHANGESET	(0)
ZCONTEXTUALCHANGEREGISTRATI...	(109)	ZCONTEXTUALCHANGEREGISTRATI...	(109)
ZCONTEXTUALKEYPATH	(0)	ZCONTEXTUALKEYPATH	(0)
ZCUSTOMMETADATA	(2205)	ZCUSTOMMETADATA	(2298)
ZCUSTOMMETADATA_ZOBJECT_I...	(2205)	ZCUSTOMMETADATA_ZOBJECT_I...	(2298)
ZDELETIONCHANGESET	(0)	ZDELETIONCHANGESET	(0)
ZHISTOGRAM	(7)	ZHISTOGRAM	(7)
ZHISTOGRAMVALUE	(355)	ZHISTOGRAMVALUE	(358)
ZHISTOGRAMVALUE_ZHISTOGRAM...	(355)	ZHISTOGRAMVALUE_ZHISTOGRAM...	(358)
ZKEYVALUE	(14)	ZKEYVALUE	(14)
ZOBJECT	(40731)	ZOBJECT	(39364)

Fig. 6.22 iPhone X comparison of ZOBJECT table from the knowledgeC.db database

In comparing the entries in the ZOBJECTS tables from both extractions against each other, the removal of the 130 Application Usage Log entries was able to be confirmed. Using a date filter of X in the X column, the entries were confirmed to have been removed between entry numbers 18,207 and 19,876.

With one entry showing an associated source file of the knowledgeC.db-wal write ahead log, the built-in hex viewer of the forensic analysis software was used to review the knowledgeC.db-wal file from both extractions. Navigating to the starting offset for the log entry of $0 \times 0275D83$ in each write ahead log, confirmed the removal of the entry, as the area in the 72 h extraction had been overwritten with 0×00, shown in Fig. 6.23.

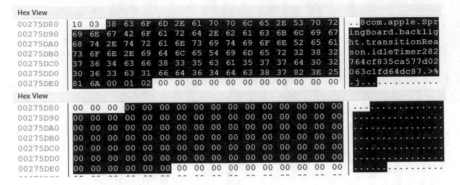

Fig. 6.23 iPhone X comparison of offset 0 × 0275D83 in the knowledgeC.db-wal 24 h versus 72 h

<u>Log Entries</u>: The artifacts of interest comparison showed a decrease in 36 entries. The log entries from the 24 h and 72 h extractions were compared, and 36 entries were identified in the 24 h extraction that were not in the 72 h extraction. These entries were then tagged in the 24 h extraction and reviewed. These 36 entries are shown in Fig. 6.24.

Out of the 36 entries, 12 were associated to the interactionC.db (ZINTERACTIONS table) database, and the remaining 24 were associated to the DataUsage.sqlite

#	Application	Body	Date	Source file information
1	com.apple.datausage.location	Wifi In:0Wifi Out:0Wan In:8231Wan Out:6608	2021-05-29	DataUsage.sqlite : 0x18F67 (Table: ZLIVEUSAGE, ZPROCESS)
2	com.apple.datausage.general	Wifi In:0Wifi Out:0Wan In:81482Wan Out:32407	2021-05-30	DataUsage.sqlite : 0x17128 (Table: ZLIVEUSAGE, ZPROCESS)
3	com.apple.datausage.general	Wifi In:0Wifi Out:0Wan In:7473Wan Out:3173	2021-06-03	DataUsage.sqlite : 0x17128 (Table: ZLIVEUSAGE, ZPROCESS)
4	com.apple.datausage.bluetooth	Wifi In:0Wifi Out:0Wan In:69813Wan Out:31341	2021-05-30	DataUsage.sqlite : 0x18150 (Table: ZLIVEUSAGE, ZPROCESS)
5	com.apple.icloud.searchpartyd	Wifi In:0Wifi Out:0Wan In:66240Wan Out:33491	2021-05-20	DataUsage.sqlite : 0x18CE6 (Table: ZLIVEUSAGE, ZPROCESS)
6	com.apple.datausage.diagnostics	Wifi In:0Wifi Out:0Wan In:6607Wan Out:2794	2021-06-03	DataUsage.sqlite : 0x18908 (Table: ZLIVEUSAGE,ZPROCESS)
7	com.apple.datausage.siri	Wifi In:0Wifi Out:0Wan In:615605Wan Out:1546535	2021-06-02	DataUsage.sqlite : 0x171D8 (Table: ZLIVEUSAGE, ZPROCESS)
8	com.apple.datausage.bluetooth	Wifi In:0Wifi Out:0Wan In:5844Wan Out:2768	2021-06-03	DataUsage.sqlite : 0x18150 (Table: ZLIVEUSAGE, ZPROCESS)
9	FaceTime	Wifi In:0Wifi Out:0Wan In:5632Wan Out:5632	2021-06-02	DataUsage.sqlite : 0x180C2 (Table: ZLIVEUSAGE, ZPROCESS)
10	com.apple.datausage.dns	Wifi In:0Wifi Out:0Wan In:519Wan Out:1883	2021-06-03	DataUsage.sqlite : 0x1717F (Table: ZLIVEUSAGE, ZPROCESS)
11	FaceTime	Wifi In:0Wifi Out:0Wan In:49308148Wan Out:63044317	2021-05-29	DataUsage.sqlite : 0x1828D (Table: ZLIVEUSAGE, ZPROCESS)
12	com.apple.datausage.dns	Wifi In:0Wifi Out:0Wan In:367232Wan Out:318659	2021-06-02	DataUsage.sqlite : 0x1717F (Table: ZLIVEUSAGE, ZPROCESS)
13	com.apple.datausage.appleid	Wifi In:0Wifi Out:0Wan In:356015Wan Out:44380	2021-06-03	DataUsage.sqlite : 0x1843B (Table: ZLIVEUSAGE, ZPROCESS)
14	com.apple.datausage.general	Wifi In:0Wifi Out:0Wan In:3521Wan Out:2106	2021-06-03	DataUsage.sqlite : 0x17147 (Table: ZLIVEUSAGE, ZPROCESS)
15	com.apple.datausage.media	Wifi In:0Wifi Out:0Wan In:3388Wan Out:8712	2021-05-29	DataUsage.sqlite : 0x1874A (Table: ZLIVEUSAGE, ZPROCESS)
16	com.apple.datausage.applepushservice	Wifi In:0Wifi Out:0Wan In:33691Wan Out:88867	2021-06-03	DataUsage.sqlite : 0x181B2 (Table: ZLIVEUSAGE, ZPROCESS)
17	com.apple.datausage.siri	Wifi In:0Wifi Out:0Wan In:254Wan Out:20925	2021-06-03	DataUsage.sqlite : 0x171D8 (Table: ZLIVEUSAGE, ZPROCESS)
18	com.apple.datausage.applepushservice	Wifi In:0Wifi Out:0Wan In:2147538Wan Out:3670981	2021-06-02	DataUsage.sqlite : 0x181B2 (Table: ZLIVEUSAGE, ZPROCESS)
19	com.apple.datausage.iad	Wifi In:0Wifi Out:0Wan In:136756Wan Out:49421	2021-06-03	DataUsage.sqlite : 0x19CB7 (Table: ZLIVEUSAGE, ZPROCESS)
20	com.apple.datausage.security	Wifi In:0Wifi Out:0Wan In:11960Wan Out:8044	2021-06-02	DataUsage.sqlite : 0x18A19 (Table: ZLIVEUSAGE, ZPROCESS)
21	com.apple.datausage.location	Wifi In:0Wifi Out:0Wan In:11807002Wan Out:830321	2021-06-02	DataUsage.sqlite : 0x1821C (Table: ZLIVEUSAGE, ZPROCESS)
22	com.apple.datausage.appleid	Wifi In:0Wifi Out:0Wan In:11532703Wan Out:5228070	2021-05-29	DataUsage.sqlite : 0x1843B (Table: ZLIVEUSAGE, ZPROCESS)
23	com.apple.datausage.icloud	Wifi In:0Wifi Out:0Wan In:11441Wan Out:4519	2021-05-16	DataUsage.sqlite : 0x1837D (Table: ZLIVEUSAGE, ZPROCESS)
24	com.apple.datausage.maps	Wifi In:0Wifi Out:0Wan In:1045115Wan Out:421977	2021-05-29	DataUsage.sqlite : 0x180FE (Table: ZLIVEUSAGE, ZPROCESS)
25	com.apple.InCallService	outgoing call	2021-05-21	interactionC.db : 0xD4D64 (Table: ZINTERACTIONS)
26	com.apple.InCallService	outgoing call	2021-05-21	interactionC.db : 0xD4DE9 (Table: ZINTERACTIONS)
27	com.apple.InCallService	outgoing call	2021-05-21	interactionC.db : 0xD4F18 (Table: ZINTERACTIONS)
28	com.apple.InCallService	outgoing call	2021-05-21	interactionC.db : 0xCB4D2 (Table: ZINTERACTIONS)
29	com.apple.InCallService	outgoing call	2021-05-20	interactionC.db : 0xAEEFC (Table: ZINTERACTIONS)
30	com.apple.InCallService	outgoing call	2021-05-20	interactionC.db : 0xAB25A (Table: ZINTERACTIONS)
31	com.apple.InCallService	outgoing call	2021-05-20	interactionC.db : 0xAB363 (Table: ZINTERACTIONS)
32	com.apple.InCallService	outgoing call	2021-05-20	interactionC.db : 0xAB96E (Table: ZINTERACTIONS)
33	com.apple.InCallService	incoming call	2021-05-21	interactionC.db : 0xD4FA8 (Table: ZINTERACTIONS)
34	com.apple.InCallService	incoming call	2021-05-20	interactionC.db : 0xAEE77 (Table: ZINTERACTIONS)
35	com.apple.InCallService	incoming call	2021-05-20	interactionC.db : 0xAB2DE (Table: ZINTERACTIONS)
36	com.apple.InCallService	incoming call	2021-05-20	interactionC.db : 0xAB3F3 (Table: ZINTERACTIONS)

Fig. 6.24 iPhone X Log lost log entries 24 h extraction

Name	Size	Modified	Name	Size	Modified
interactionC.db	1 572 864	2021-06-11 16:48	interactionC.db	1 572 864	2021-06-11 16:48
interactionC.db-shm	32 768	2021-06-16 17:39	interactionC.db-shm	32 768	2021-06-18 17:39
interactionC.db-wal	3 427 872	2021-06-16 17:39	interactionC.db-wal	3 843 992	2021-06-18 17:39

Fig. 6.25 iPhone X interactionC.db database files comparison 24 h versus 72 h

(ZLIVEUSAGE, ZPROCESS table). Further analysis began with the 12 entries associated to the interactionC.db database first. Each of the 12 entries were noted to be associated to the com.apple.InCallService application.

When comparing the interactionC.db from the iPhone X's 24 h and 72 h extractions, the interactionC.db did not change in size, however; a new modified time of 2021-06-18 17:39 was noted in 72 h extraction, shown in Fig. 6.25.

When viewing the ZINTERACTIONS table of the interactionC.db database from both extractions, a reduction in the number of table entries from 2393 to 2381 was observed, seen in Fig. 6.26.

In comparing the entries from both ZINTERACTIONS tables, the 12 entries were confirmed to have been removed. Using a date filter of 2021-05-20 to 2021-05-21 on the ZENDATE column, and a filter of com.apple.InCallService on the ZBUNDLEID column, the 12 log entries were identified as table entry numbers 1113, 1122–1125, 1130, 1131, 1432, 1493, 1494, 1496 and 1497, shown in Fig. 6.27.

With the removal of the 12 entries from the interactionC.db confirmed, analy-sis was conducted on the remaining 24 log entries associated to the DataU-sage.sqlite database.

When comparing the DataUsage.sqlite database files from the iPhone X's 24 h and 72 h extractions, a new modified time of 2021-06-18 17:34 as well an in-crease in size of 568,560 B was noted in the 72 h extraction for the DataU-sage.sqlite-wal write ahead log, shown in Fig. 6.28.

Comparing the ZLIVEUSAGE tables in the DataUsage.sqlite database from both extractions a reduction in entries from 258 to 234 was observed, shown in Fig. 6.29.

In comparing the entries in the ZLIVEUSAGE tables from both extractions, the 24 entries were confirmed to have been removed. Using a filter on the ZTIMESTAMP column of the 2021-05-16 to 2021-06-03, the removed were confirmed to be entry numbers 471–472, 477, 483, 491–492, 496–499, 783, 785, 788, 793, 795–796, 801 and 803, 819–802, 822, 825, 828, and 849. These entries can be seen in Fig. 6.30.

Images: A decrease of 287 images was identified in the comparison of artifacts of interest from the iPhone X's 24 h and 72 h extractions. The images from both extractions were reviewed, and the 287 images were identified in the 24 h extraction that were not in the 72 h extraction. These images were tagged and reviewed.

Of the 287 images, 55 were full size photos located in the iPhone X's file system at /private/var/mobile/Media/DCIM/100APPLE/ folder. Each of these 55 images were marked as being in the Trash. Comparing the 100APPLE folders from the iPhone X's 24 h and 72 h extractions showed a decrease in folder size of 94,679,249 B and a new modified time in the 72 h extraction of 2021-06-18 05:37, depicted in Fig. 6.31.

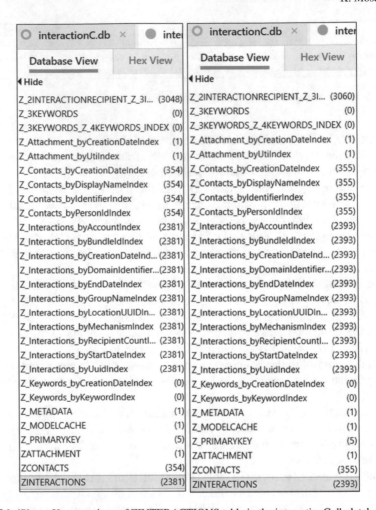

Fig. 6.26 iPhone X comparison of ZINTERACTIONS table in the interactionC.db database

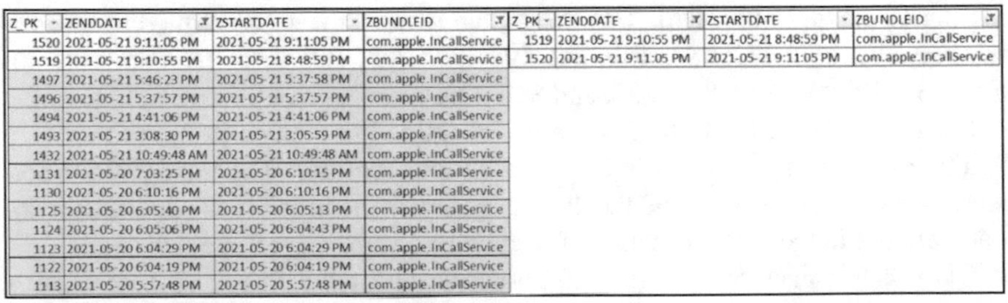

Fig. 6.27 iPhone X comparison of the ZINTERACTIONS table entries 24 h versus 72 h

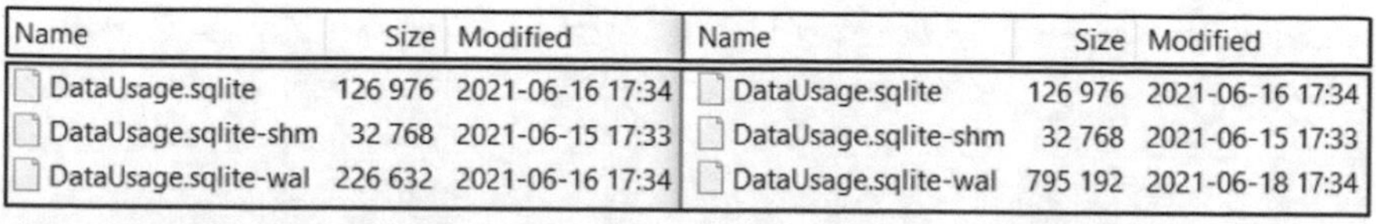

Name	Size	Modified	Name	Size	Modified
DataUsage.sqlite	126 976	2021-06-16 17:34	DataUsage.sqlite	126 976	2021-06-16 17:34
DataUsage.sqlite-shm	32 768	2021-06-15 17:33	DataUsage.sqlite-shm	32 768	2021-06-15 17:33
DataUsage.sqlite-wal	226 632	2021-06-16 17:34	DataUsage.sqlite-wal	795 192	2021-06-18 17:34

Fig. 6.28 iPhone X comparison of the DataUsage.sqlite database files 24 h versus 72 h

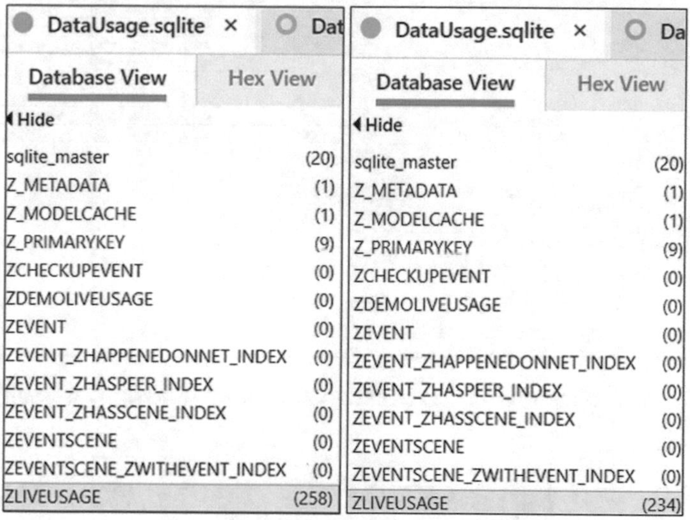

Fig. 6.29 iPhone X comparison of ZLIVEUSAGE tables in the DataUsage.sqlite database

In reviewing the /private/var/mobile/Media/DCIM/100APPLE/ folder, the removal of the 55 images was confirmed. A tally of the contents of this folder from the 24 h and 72 h extractions can be seen in Fig. 6.32.

Unlike the previous 55 images, the remaining 232 images were all thumbnail images, meaning small versions (likened to the size of a thumbnail) of a larger image files. These 232 thumbnails were spread across five different locations from within the /private/var/mobile/media/PhotoData path in the iPhone X's file system as follows:

- 4 located in /private/var/mobile/Media/PhotoData/Metadata/DCIM/100APPLE/ (which is the video thumbnails folder);
- 57 located in /private/var/mobile/Media/PhotoData/Thumbnails/V2/DCIM/100APPLE/ (which is the photo thumbnails folder); and
- 171 located in /private/var/mobile/Media/PhotoData/Thumbnails/.

Located within the /private/var/mobile/Media/PhotoData/Thumbnails/ folder, were three numbered '.ithmb' data files. These numbered data files store picture and video thumbnails in '.bmp' format based on resolution size. The thumbnail images stored in each of these data files are all the same size. The first data file,

Z_PK	ZHASPROCESS	ZTIMESTAMP	ZWWANIN	ZWWANOUT	Z_PK	ZHASPROCESS	ZTIMESTAMP	ZWWANIN	ZWWANOUT
443	36	2021-06-02 21:18	903	257	443	36	2021-06-02 21:18	903	257
471	73	2021-06-03 6:44	5844	2768	490	112	2021-06-03 17:31	47742	18032
472	39	2021-06-03 6:44	7473	3173	815	102	2021-05-29 20:19	22180	5864
477	60	2021-06-03 10:59	11960	8044	817	104	2021-05-30 15:49	1678217	165760
483	52	2021-05-20 9:26	66240	33491	824	50	2021-06-02 21:18	486440	206677
491	86	2021-06-03 17:50	3521	2106	829	74	2021-06-02 14:55	11571	22604
492	63	2021-06-03 17:50	6607	2794	830	75	2021-06-02 14:55	21211298	1028653
496	43	2021-06-03 21:03	33691	88867	835	88	2021-06-02 21:18	268257	105736
497	36	2021-06-03 22:34	519	1883	836	90	2021-06-02 21:18	5785	7221
498	80	2021-06-03 23:46	356015	44380	838	93	2021-06-02 21:18	126574	47606
499	38	2021-06-04 2:32	254	20925	839	94	2021-06-02 21:18	192172	188166
783	39	2021-05-30 6:00	81482	32407	841	97	2021-06-02 21:18	96139	31671
785	42	2021-05-29 20:19	8231	6608	842	98	2021-06-02 21:18	83942	46392
788	49	2021-05-29 20:19	1045115	421977	843	99	2021-06-02 21:18	464718	144279
793	68	2021-05-29 22:45	3388	8712	847	105	2021-06-02 21:18	61728	17713
795	70	2021-05-29 14:35	49308148	63044317	853	42	2021-06-03 10:46	10682	8480
796	73	2021-05-30 6:00	69813	31341	855	49	2021-06-03 10:17	1116635	452533
801	80	2021-05-29 21:48	11532703	5228070	856	52	2021-06-03 10:46	98138	49612
803	82	2021-05-17 2:39	11441	4519	861	70	2021-06-03 21:01	51230159	67427556
815	102	2021-05-29 20:19	22180	5864	864	82	2021-06-03 10:56	37504	20146
819	36	2021-06-02 21:18	367232	318659					
820	38	2021-06-03 1:26	615605	1546535					
822	43	2021-06-02 18:38	2147538	3670981					
824	50	2021-06-02 21:18	486440	206677					
825	51	2021-06-02 21:18	11807002	830321					
828	69	2021-06-02 13:14	5632	5632					
835	88	2021-06-02 21:18	268257	105736					
836	90	2021-06-02 21:18	5785	7221					
838	93	2021-06-02 21:18	126574	47606					
839	94	2021-06-02 21:18	192172	188166					
840	96	2021-06-02 21:18	136756	49421					
841	97	2021-06-02 21:18	96139	31671					

Fig. 6.30 iPhone X comparison of the ZLIVEUSAGE entries 24 h versus 72 h identifying the 24 lost entries

Name	Size	Modified	Name	Size	Modified
100APPLE	2 462 278 306	2021-06-15 18:34	100APPLE	2 367 599 057	2021-06-18 05:37

Fig. 6.31 iPhone X DCIM/100APPLE folder comparison 24 h versus 72 h

Fig. 6.32 iPhone X /private/var/mobile/Library/DCIM/100APPLE folder comparison 24 h versus 72 h

3306.ithmb, contained 57 small thumbnails, all 3186 B in size. The second data file, 3314.ithmb, contained 57 cropped thumbnails, each one 31,566 B in size. The third data file, 4031.ithmb, contained 57 full size thumbnails, each one 28,866 B in size.

In analysing /private/var/mobile/Media/PhotoData/Metadata/DCIM/100APPLE and /private/var/mobile/Media/PhotoData/Thumbnails/V2/DCIM/100APPLE/, the removal of the 61 photo thumbnail images, seen in Fig. 6.33.

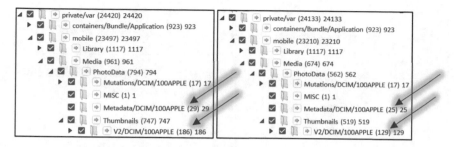

Fig. 6.33 iPhone X file system comparison 24 h versus 72 h showing the 61 lost thumbnail images

Additionally, in viewing the three numbered data files, the removal of the 171 images was confirmed. The contents of each data file were extracted from the 24 h and 72 h extractions. A view of the folder properties containing the embedded image files from the 3306.ithmb data file is shown as an example in Figure below. The removed images from the 3306.ithmb, 3314.ithmb and 4031.ithmb are shown in Figs. 6.34, 6.35 and 6.36.

In comparing all five locations where the combined 232 thumbnail images were stored (2 folder paths and 3 data files) a common modified time of 2021-06-18 05:37 was noted for all locations in the 72 h extraction. Also noted, was that only the folder locations saw a decrease in size while the '.ithmb' data file sizes remained the same. A comparison of two folders and three files can be seen in Fig. 6.37.

Videos: Comparing the artifacts of interest from the 24 h and 72 h extraction identified a decrease of 57 videos. The artifacts from both extractions were compared, and 57 videos were identified in the 24 h extraction that were not found in the 72 h extraction, which were tagged and reviewed. Each of the videos had a source folder location of /private/var/mobile/Media/DCIM/100APPLE/ and two of the 57 were identified

	3306.ithmb.EMBEDDED			3306.ithmb.EMBEDDED
Type:	File folder (.EMBEDDED)		Type:	File folder (.EMBEDDED)
Location:			Location:	
Size:	578 KB (592,596 bytes)		Size:	578 KB (592,596 bytes)
Size on disk:	744 KB (761,856 bytes)		Size on disk:	744 KB (761,856 bytes)
Contains:	186 Files, 0 Folders		Contains:	186 Files, 0 Folders

Fig. 6.34 iPhone X comparison of 3306.ithmb properties 24 h versus 72 h showing 57 removed files

3314.ithmb.EMBEDDED
Type: File folder (.EMBEDDED)
Location: H:\iPhone X June 15 1734 h
Size: 5.59 MB (5,871,276 bytes)
Size on disk: 5.81 MB (6,094,848 bytes)
Contains: 186 Files, 0 Folders
3314.ithmb.EMBEDDED
Type: File folder (.EMBEDDED)
Location:
Size: 3.88 MB (4,072,014 bytes)
Size on disk: 4.03 MB (4,227,072 bytes)
Contains: 129 Files, 0 Folders

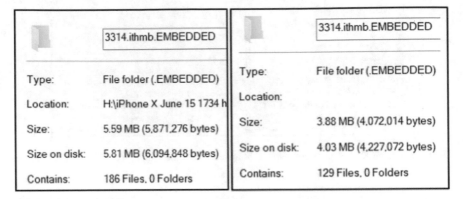

Fig. 6.35 iPhone X comparison of 3314.ithmb properties 24 h versus 72 h showing 57 removed files

4031.ithmb.EMBEDDED
Type: File folder (.EMBEDDED)
Location: H:\iPhone X June 15 1734 h
Size: 3.55 MB (3,723,714 bytes)
Size on disk: 4.03 MB (4,227,072 bytes)
Contains: 129 Files, 0 Folders
4031.ithmb.EMBEDDED
Type: File folder (.EMBEDDED)
Location: H:\iPhone X June 15 1734 h
Size: 3.55 MB (3,723,714 bytes)
Size on disk: 4.03 MB (4,227,072 bytes)
Contains: 129 Files, 0 Folders

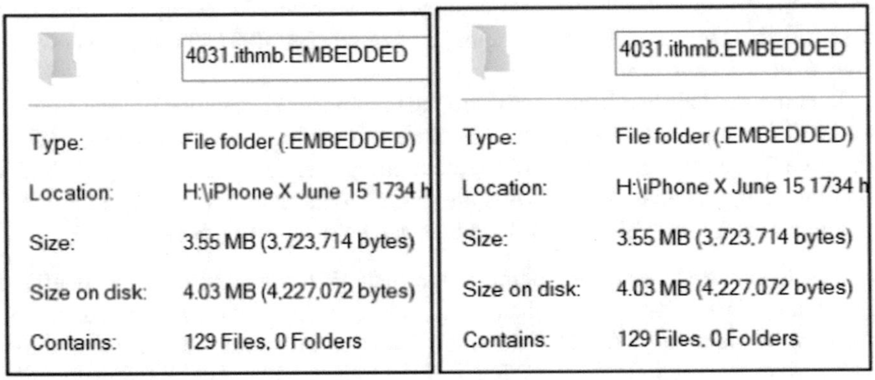

Fig. 6.36 iPhone X comparison of 4031.ithmb properties 24 h versus 72 h showing 57 removed files

Name Size Modified
100APPLE 2 208 464 2021-06-09 17:38
100APPLE 8 834 617 2021-06-15 18:34
3306.ithmb 604 416 2021-06-15 18:34
3314.ithmb 6 149 376 2021-06-15 18:34
4031.ithmb 5 534 976 2021-06-15 18:34
Name Size Modified
100APPLE 1 728 507 2021-06-18 05:37
100APPLE 6 505 036 2021-06-18 05:37
3306.ithmb 604 416 2021-06-18 05:37
3314.ithmb 6 149 376 2021-06-18 05:37
4031.ithmb 5 534 976 2021-06-18 05:37

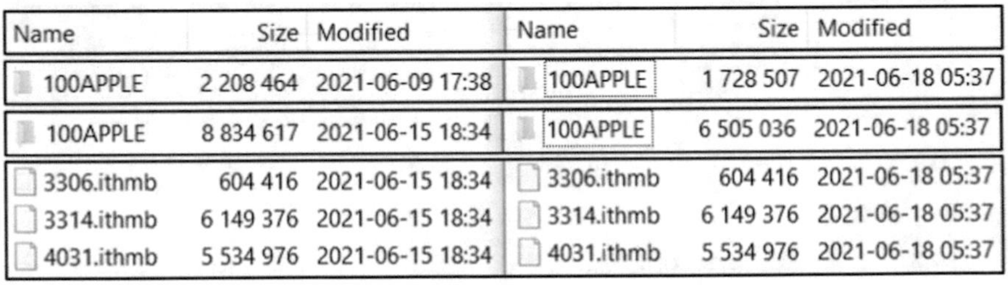

Fig. 6.37 iPhone X comparison of thumbnail source locations 24 h versus 72 h

#	Created-Time (UTC-3)	Deleted	Name and Size	#	Created-Time (UTC-3)	Deleted	Name and Size
			Source: /private/var/mobile/Media/DCIM/100APPLE/				
1	2021-05-18 13:00	Trash	IMG_0070.MOV (Size: 30425936 bytes)	30	2021-05-14 13:44		IMG_0039.MOV (Size: 184011 bytes)
2	2021-05-15 17:59		IMG_0067.MOV (Size: 1853934 bytes)	31	2021-05-14 13:44		IMG_0038.MOV (Size: 184011 bytes)
3	2021-05-15 17:59		IMG_0066.MOV (Size: 1853934 bytes)	32	2021-05-14 13:44		IMG_0037.MOV (Size: 184011 bytes)
4	2021-05-15 17:59		IMG_0065.MOV (Size: 1850393 bytes)	33	2021-05-14 13:44		IMG_0036.MOV (Size: 184011 bytes)
5	2021-05-15 17:59		IMG_0064.MOV (Size: 1850393 bytes)	34	2021-05-14 13:44		IMG_0035.MOV (Size: 220116 bytes)
6	2021-05-15 17:59		IMG_0063.MOV (Size: 1738969 bytes)	35	2021-05-14 13:44		IMG_0034.MOV (Size: 220116 bytes)
7	2021-05-15 17:59		IMG_0062.MOV (Size: 1818373 bytes)	36	2021-05-14 13:44		IMG_0033.MOV (Size: 220116 bytes)
8	2021-05-14 18:42		IMG_0061.MOV (Size: 330945 bytes)	37	2021-05-14 13:44		IMG_0032.MOV (Size: 220116 bytes)
9	2021-05-14 13:44		IMG_0060.MOV (Size: 178304 bytes)	38	2021-05-14 13:44		IMG_0031.MOV (Size: 220116 bytes)
10	2021-05-14 13:44		IMG_0059.MOV (Size: 178304 bytes)	39	2021-05-14 13:44		IMG_0030.MOV (Size: 220116 bytes)
11	2021-05-14 13:44		IMG_0058.MOV (Size: 178304 bytes)	40	2021-05-14 13:44		IMG_0029.MOV (Size: 663500 bytes)
12	2021-05-14 13:44		IMG_0057.MOV (Size: 178304 bytes)	41	2021-05-14 13:44		IMG_0028.MOV (Size: 681674 bytes)
13	2021-05-14 13:44		IMG_0056.MOV (Size: 178304 bytes)	42	2021-05-14 13:44		IMG_0027.MOV (Size: 654896 bytes)
14	2021-05-14 13:44		IMG_0055.MOV (Size: 178304 bytes)	43	2021-05-14 11:15		IMG_0026.MOV (Size: 1858186 bytes)
15	2021-05-14 13:44		IMG_0054.MOV (Size: 178304 bytes)	44	2021-05-14 17:14		IMG_0024.MOV (Size: 1711351 bytes)
16	2021-05-14 13:44		IMG_0053.MOV (Size: 178304 bytes)	45	2021-05-13 17:14		IMG_0023.MOV (Size: 1711351 bytes)
17	2021-05-14 13:44		IMG_0052.MOV (Size: 178304 bytes)	46	2021-05-13 17:14		IMG_0022.MOV (Size: 1711351 bytes)
18	2021-05-14 13:44		IMG_0051.MOV (Size: 178304 bytes)	47	2021-05-13 17:14		IMG_0021.MOV (Size: 1711351 bytes)
19	2021-05-14 13:44		IMG_0050.MOV (Size: 178304 bytes)	48	2021-05-13 17:14		IMG_0020.MOV (Size: 1711351 bytes)
20	2021-05-14 13:44		IMG_0049.MOV (Size: 178304 bytes)	49	2021-05-13 17:14		IMG_0019.MOV (Size: 1711351 bytes)
21	2021-05-14 13:44		IMG_0048.MOV (Size: 184011 bytes)	50	2021-05-13 17:14		IMG_0018.MOV (Size: 1711351 bytes)
22	2021-05-14 13:44		IMG_0047.MOV (Size: 184011 bytes)	51	2021-05-13 17:14		IMG_0017.MOV (Size: 1711351 bytes)
23	2021-05-14 13:44		IMG_0046.MOV (Size: 184011 bytes)	52	2021-05-13 17:14		IMG_0016.MOV (Size: 1711351 bytes)
24	2021-05-14 13:44		IMG_0045.MOV (Size: 184011 bytes)	53	2021-05-13 17:14		IMG_0015.MOV (Size: 1711351 bytes)
25	2021-05-14 13:44		IMG_0044.MOV (Size: 184011 bytes)	54	2021-05-13 17:14		IMG_0014.MOV (Size: 1711351 bytes)
26	2021-05-14 13:44		IMG_0043.MOV (Size: 184011 bytes)	55	2021-05-13 11:03		IMG_0013.MOV (Size: 2350464 bytes)
27	2021-05-14 13:44		IMG_0042.MOV (Size: 184011 bytes)	56	2021-05-13 9:23	Trash	IMG_0012.MOV (Size: 6718699 bytes)
28	2021-05-14 13:44		IMG_0041.MOV (Size: 184011 bytes)	57	2021-05-13 9:23		IMG_0011.MOV (Size: 2584513 bytes)
29	2021-05-14 13:44		IMG_0040.MOV (Size: 184011 bytes)				

Fig. 6.38 iPhone X 57 lost video files 72 h extraction

as being in the Trash by the forensic analysis software. The list of 57 videos can be seen in Fig. 6.38.

The source folder location for these 57 videos is the same location as the 55 full size lost images identified above. Therefore, the decrease in folder size and change in modified time can be seen in Fig. 6.38.

In reviewing each of the folder locations that contained the 57 videos in the iPhone X's file system, the removal of each video was confirmed, shown in Fig. 6.39.

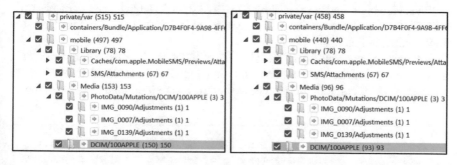

Fig. 6.39 iPhone X file system comparison 24 h versus 72 h showing removal of 57 videos

72 h versus 7 days

Application Usage Log: A decrease of 615 Application Usage Log entries was identified in the artifacts of interest comparison between iPhone X's 72 h and 7 day extractions. In reviewing both extractions, the 615 entries were identified in the 72 h extraction, tagged and reviewed. Of the 615 entries, 275 have an associated source file of the knowledgeC.db (ZOBJECT, ZSTRUCTUREMETADATA tables) database and the remaining 340 have an associated source file of the knowledgeC.db (ZOBJECT table) database.

Comparing the knowledgeC.db databases from the iPhone X's 72 h and 7 day extractions, a new modified time of 2021-06-22 17:44 as well as a decrease in size of 6,864,496 B for the knowledgeC.db database was noted in the 72 h extraction. These changes can be seen in Fig. 6.40.

In reviewing the ZOBJECTS table in knowledgeC.db database from the 72 h and 7 day extraction, a reduction in entries from 39,364 to 31,858 was noted, shown in Fig. 6.41.

In reviewing the entries in the ZOBJECT table from both extraction, the 615 Application Usage Log entries were confirmed to have been removed between entry numbers 20,046 and 27,976.

Log Entries: A decrease of 87 entries was identified in the artifacts of interest comparison. In reviewing the log entries from the 72 h and 7 days extraction, a decrease of 88 entries was identified in the 72 h extraction and one new log entry was identified in the 7 days extraction. All 89 changes were tagged and reviewed. Of the 88 entries, 17 had an associated source file of the interactionC.db (ZINTERACTIONS table) database, and all 17 entries were associated to the application com.apple.InCallService. A listing of the 17 entries can be seen in Fig. 6.42.

When comparing the interactionC.db database from the iPhone X's 72 h and 7 day extractions, a new modified time of 2021-06-20 13:39 and an increase in size of 12,288 B was noted for the interactionC.db in the 7 day extraction, seen in Fig. 6.43.

When viewing the contents of the ZINTERACTIONS table in the interactionC.db database from both extractions, a decrease in table entries from 2381 to 2364 was noted, shown in Fig. 6.44.

In comparing the contents of the ZINTERACTIONS tables from both extraction, the 17 removed entries were confirmed. Using a filter of 2021-05-21 to 2021-05-25 in the ZENDATE column, the 17 removed entries were identified as entry numbers

Name	Size	Modified	Name	Size	Modified
knowledgeC.db	41 119 744	2021-06-18 17:37	knowledgeC.db	34 254 848	2021-06-22 17:44
knowledgeC.db-shm	32 768	2021-06-15 17:34	knowledgeC.db-shm	32 768	2021-06-15 17:34
knowledgeC.db-wal	4 115 912	2021-06-18 17:46	knowledgeC.db-wal	774 592	2021-06-22 17:46

Fig. 6.40 iPhone X knowledgeC.db database files comparison 72 h versus 7 days

knowledgeC.db — Database View		knowledgeC.db — Database View	
Z_Event_compoundIndex	(39364)	Z_Event_compoundIndex	(31858)
Z_Event_compoundIndex1	(39364)	Z_Event_compoundIndex1	(31858)
Z_KeyValue_UNIQUE_domain_key	(14)	Z_KeyValue_UNIQUE_domain_key	(14)
Z_METADATA	(1)	Z_METADATA	(1)
Z_MODELCACHE	(1)	Z_MODELCACHE	(1)
Z_Object_UNIQUE_uuid	(39364)	Z_Object_UNIQUE_uuid	(31858)
Z_Object_uuidIndex	(39364)	Z_Object_uuidIndex	(31858)
Z_PRIMARYKEY	(17)	Z_PRIMARYKEY	(17)
Z_StructuredMetadata_UNIQUE_...	(6457)	Z_StructuredMetadata_UNIQUE_...	(4729)
Z_SyncPeer_UNIQUE_uuid	(1)	Z_SyncPeer_UNIQUE_uuid	(1)
ZADDITIONCHANGESET	(0)	ZADDITIONCHANGESET	(0)
ZCONTEXTUALCHANGEREGISTRATI...	(109)	ZCONTEXTUALCHANGEREGISTRATI...	(109)
ZCONTEXTUALKEYPATH	(0)	ZCONTEXTUALKEYPATH	(0)
ZCUSTOMMETADATA	(2298)	ZCUSTOMMETADATA	(2369)
ZCUSTOMMETADATA_ZOBJECT_I...	(2298)	ZCUSTOMMETADATA_ZOBJECT_I...	(2369)
ZDELETIONCHANGESET	(0)	ZDELETIONCHANGESET	(0)
ZHISTOGRAM	(7)	ZHISTOGRAM	(7)
ZHISTOGRAMVALUE	(358)	ZHISTOGRAMVALUE	(392)
ZHISTOGRAMVALUE_ZHISTOGRAM...	(358)	ZHISTOGRAMVALUE_ZHISTOGRAM...	(392)
ZKEYVALUE	(14)	ZKEYVALUE	(14)
ZOBJECT	(39364)	ZOBJECT	(31858)

Fig. 6.41 iPhone X ZOBJECT table from the knowledgeC.db database comparison 72 h versus 7 days

#	Application	Body	Start Time (UTC-3)	End time (UTC-3)	Source file information
68	com.apple.InCallService	incoming call	2021-05-25 15:42	2021-05-25 16:14	interactionC.db : 0x12D13B (Table: ZINTERACTIONS)
69	com.apple.InCallService	outgoing call	2021-05-25 15:41	2021-05-25 15:41	interactionC.db : 0x12D1C0 (Table: ZINTERACTIONS)
70	com.apple.InCallService	incoming call	2021-05-25 12:09	2021-05-25 12:13	interactionC.db : 0x12D250 (Table: ZINTERACTIONS)
71	com.apple.InCallService	outgoing call	2021-05-25 10:15	2021-05-25 10:15	interactionC.db : 0x12D373 (Table: ZINTERACTIONS)
72	com.apple.InCallService	outgoing call	2021-05-24 12:55	2021-05-24 13:00	interactionC.db : 0x12C9F0 (Table: ZINTERACTIONS)
73	com.apple.InCallService	outgoing call	2021-05-23 18:49	2021-05-23 19:29	interactionC.db : 0xF6AC0 (Table: ZINTERACTIONS)
74	com.apple.InCallService	outgoing call	2021-05-23 18:45	2021-05-23 18:49	interactionC.db : 0xF6B44 (Table: ZINTERACTIONS)
75	com.apple.InCallService	incoming call	2021-05-23 10:48	2021-05-23 11:21	interactionC.db : 0xF6BC8 (Table: ZINTERACTIONS)
76	com.apple.InCallService	outgoing call	2021-05-23 9:12	2021-05-23 10:03	interactionC.db : 0xF6C4C (Table: ZINTERACTIONS)
77	com.apple.InCallService	outgoing call	2021-05-23 9:11	2021-05-23 9:12	interactionC.db : 0xF6CD0 (Table: ZINTERACTIONS)
78	com.apple.InCallService	incoming call	2021-05-22 17:18	2021-05-22 19:34	interactionC.db : 0xDEDBC (Table: ZINTERACTIONS)
79	com.apple.InCallService	outgoing call	2021-05-22 17:17	2021-05-22 17:17	interactionC.db : 0xDD851 (Table: ZINTERACTIONS)
80	com.apple.InCallService	outgoing call	2021-05-22 17:14	2021-05-22 17:14	interactionC.db : 0xDD8E2 (Table: ZINTERACTIONS)
81	com.apple.InCallService	incoming call	2021-05-22 16:06	2021-05-22 17:02	interactionC.db : 0xDD972 (Table: ZINTERACTIONS)
82	com.apple.InCallService	outgoing call	2021-05-22 14:28	2021-05-22 14:28	interactionC.db : 0xD582A (Table: ZINTERACTIONS)
86	com.apple.InCallService	outgoing call	2021-05-21 18:11	2021-05-21 18:11	interactionC.db : 0xD5DCA (Table: ZINTERACTIONS)
87	com.apple.InCallService	outgoing call	2021-05-21 17:48	2021-05-21 18:10	interactionC.db : 0xD5E5A (Table: ZINTERACTIONS)

Fig. 6.42 iPhone X 17 log entries removed from the 72 h extraction

Name	Size	Modified	Name	Size	Modified
interactionC.db	1 572 864	2021-06-11 16:48	interactionC.db	1 585 152	2021-06-20 13:39
interactionC.db-shm	32 768	2021-06-18 17:39	interactionC.db-shm	32 768	2021-06-22 17:39
interactionC.db-wal	3 843 992	2021-06-18 17:39	interactionC.db-wal	432 632	2021-06-22 17:39

Fig. 6.43 iPhone X interactionC.db database files comparison 72 h versus 7 days

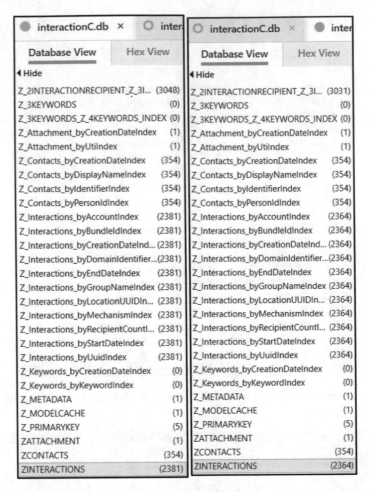

Fig. 6.44 iPhone X ZINTERACTIONS table from the interactionC.db database comparison 72 h versus 7 days

1519, 1520, 1536, 1587–1589, 1600, 1838, 1840, 1841, 1844, 1846, 2606, 2641, and 2643–2645, shown in Fig. 6.45.

The remaining 71 Log Entries had an associated source file of the DataUsage.sqlite-wal (ZLIVEUSAGE, ZPROCESS tables) write ahead log. A listing of these 71 log entries can be seen in Fig. 6.46.

The additional 1 new log entry in the 7 days extraction also had an associated source file of the DataUsage.sqlite-wal (ZLIVEUSAGE, ZPROCESS tables) write ahead log, shown in Fig. 6.47.

When comparing the DataUsage.sqlite database files from the iPhone X's 72 h and 7 day extractions, a new modified time of 2021-06-22 17:34 and an in-crease

Z_PK	ZENDDATE	ZSTARTDATE	ZBUNDLEID	Z_PK	ZENDDATE	ZSTARTDATE	ZBUNDLEID
2696	2021-05-25 10:59:10 PM	2021-05-25 9:36:55 PM	com.apple.InCallService	2696	2021-05-25 10:59:10 PM	2021-05-25 9:36:55 PM	com.apple.InCallService
2675	2021-05-25 9:36:23 PM	2021-05-25 9:36:23 PM	com.apple.InCallService	2675	2021-05-25 9:36:23 PM	2021-05-25 9:36:23 PM	com.apple.InCallService
2670	2021-05-25 9:35:19 PM	2021-05-25 9:34:47 PM	com.apple.InCallService	2670	2021-05-25 9:35:19 PM	2021-05-25 9:34:47 PM	com.apple.InCallService
2645	2021-05-25 7:14:05 PM	2021-05-25 6:42:18 PM	com.apple.InCallService				
2644	2021-05-25 6:41:37 PM	2021-05-25 6:41:37 PM	com.apple.InCallService				
2643	2021-05-25 3:13:03 PM	2021-05-25 3:09:08 PM	com.apple.InCallService				
2641	2021-05-25 1:15:23 PM	2021-05-25 1:15:23 PM	com.apple.InCallService				
2606	2021-05-24 4:00:55 PM	2021-05-24 3:55:44 PM	com.apple.InCallService				
1846	2021-05-23 10:29:17 PM	2021-05-23 9:49:18 PM	com.apple.InCallService				
1844	2021-05-23 9:49:13 PM	2021-05-23 9:45:14 PM	com.apple.InCallService				
1841	2021-05-23 2:21:33 PM	2021-05-23 1:48:38 PM	com.apple.InCallService				
1840	2021-05-23 1:03:44 PM	2021-05-23 12:12:37 PM	com.apple.InCallService				
1838	2021-05-23 12:12:34 PM	2021-05-23 12:11:23 PM	com.apple.InCallService				
1600	2021-05-22 10:34:39 PM	2021-05-22 8:18:49 PM	com.apple.InCallService				
1589	2021-05-22 8:17:26 PM	2021-05-22 8:17:26 PM	com.apple.InCallService				
1588	2021-05-22 8:14:15 PM	2021-05-22 8:14:15 PM	com.apple.InCallService				
1587	2021-05-22 8:02:07 PM	2021-05-22 7:06:37 PM	com.apple.InCallService				
1536	2021-05-22 5:28:57 PM	2021-05-22 5:28:57 PM	com.apple.InCallService				
1520	2021-05-21 9:11:05 PM	2021-05-21 9:11:05 PM	com.apple.InCallService				
1519	2021-05-21 9:10:55 PM	2021-05-21 8:48:59 PM	com.apple.InCallService				

Fig. 6.45 iPhone X ZINTERACTIONS table comparison 72 h versus 7 days showing 17 removed entries

in size of 1,458,480 B was noted for the DataUsage.sqlite-wal file in the 7 days extraction, shown in Fig. 6.48.

In reviewing the ZLIVEUSAGE table in the DataUsage.sqlite database from both extraction, a decrease in entries from 234 to 164 was noted, seen in Fig. 6.49.

In comparing the entries in the ZLIVEUSAGE table from both extractions, the removal of the 71 was confirmed. Using a filter of 2021-05-18 to 2021-06-11 on the ZTIMESTAMP column, the removed log entries were confirmed to be entry numbers 545–547, 557, 559, 562, 568, 575, 578, 584, 588, 589, 602–606, 629, 731, 740, 808, 815, 821, 823, 824, 826, 827, 829–833, 835–839, 841–849, 855, 859–861,864 and 866–873, shown in Fig. 6.50.

In comparing the entries in the ZLIVEUSAGE table from both extractions, using a filter in the ZTIMESTAMP column of 2021-06-11, the new entry was found in both the 72 h and 7 day extraction as entry number 721, shown in Fig. 6.51. Therefore, the entry could not be confirmed as new.

These verifications were conducted in the main DataUsage.sqlite databases, however the identified lost (and new) entries had an associated source file of the DataUsage.sqlite.

Images: The artifacts of interest comparison showed a decrease of 66 images from the iPhone X's 72 h extraction to the 7 days extraction. The images files from both extractions were reviewed and the 66 images were identified, tagged and reviewed. The 66 images had identified source locations in multiple areas within the iPhone X's file system and none were identified by the forensic software as being in the Trash or deleted.

Of the 66 images files, five had an associated source path of /private/var /mobile/media/PhotoData/. A listing of these five images is provided in Fig. 6.52.

Just like the previous lost images comparison between the iPhone X's 24 h and 72 h extractions, these file system locations store photo and video thumbnails in folders or data files. In reviewing each of these locations, the removal of all 5 thumbnails was confirmed, shown in Fig. 6.53.

#	Application	Body	Time (UTC-3)	Source file information
1	com.apple.datausage.dns	Wifi In:0Wifi Out:0Wan In:31188Wan Out:25417	2021-06-11 8:52	DataUsage.sqlite : 0xC1B8C (Table: ZLIVEUSAGE, ZPROCESS)
2	com.apple.datausage.wifiassist	Wifi In:0Wifi Out:0Wan In:6248Wan Out:3920	2021-06-08 21:07	DataUsage.sqlite : 0x9B3E5 (Table: ZLIVEUSAGE, ZPROCESS)
3	com.apple.datausage.dns	Wifi In:0Wifi Out:0Wan In:1931Wan Out:2372	2021-06-07 21:23	DataUsage.sqlite : 0x9B5CC (Table: ZLIVEUSAGE, ZPROCESS)
4	com.apple.datausage.appleid	Wifi In:0Wifi Out:0Wan In:297078Wan Out:2409348	2021-06-07 20:50	DataUsage.sqlite : 0x9B5EC (Table: ZLIVEUSAGE, ZPROCESS)
5	FaceTime	Wifi In:0Wifi Out:0Wan In:44Wan Out:44	2021-06-07 18:02	DataUsage.sqlite : 0x9B60D (Table: ZLIVEUSAGE, ZPROCESS)
6	FaceTime	Wifi In:0Wifi Out:0Wan In:5420Wan Out:4912	2021-06-07 18:02	DataUsage.sqlite : 0x9B62B (Table: ZLIVEUSAGE, ZPROCESS)
7	com.apple.datausage.applepushservice	Wifi In:0Wifi Out:0Wan In:86621Wan Out:96856	2021-06-07 17:13	DataUsage.sqlite : 0x9B64B (Table: ZLIVEUSAGE, ZPROCESS)
8	com.apple.datausage.siri	Wifi In:0Wifi Out:0Wan In:0Wan Out:2200	2021-06-07 0:46	DataUsage.sqlite : 0x9B68C (Table: ZLIVEUSAGE, ZPROCESS)
9	FaceTime	Wifi In:0Wifi Out:0Wan In:8440Wan Out:11528	2021-06-06 16:55	DataUsage.sqlite : 0x9B6A9 (Table: ZLIVEUSAGE, ZPROCESS)
10	com.apple.datausage.itunesmedia	Wifi In:0Wifi Out:0Wan In:11492Wan Out:5635	2021-06-06 14:23	DataUsage.sqlite : 0x9B725 (Table: ZLIVEUSAGE, ZPROCESS)
11	News	Wifi In:0Wifi Out:0Wan In:1407Wan Out:1280	2021-06-06 14:21	DataUsage.sqlite : 0x9B7E2 (Table: ZLIVEUSAGE, ZPROCESS)
12	com.apple.datausage.docsandsync	Wifi In:0Wifi Out:0Wan In:33153Wan Out:6783	2021-06-06 14:17	DataUsage.sqlite : 0x9B842 (Table: ZLIVEUSAGE, ZPROCESS)
13	com.apple.datausage.findmyiphone	Wifi In:0Wifi Out:0Wan In:97213Wan Out:53808	2021-06-06 14:16	DataUsage.sqlite : 0xC1190 (Table: ZLIVEUSAGE, ZPROCESS)
14	com.apple.datausage.general	Wifi In:0Wifi Out:0Wan In:59268Wan Out:40269	2021-06-06 14:16	DataUsage.sqlite : 0xC0D8A (Table: ZLIVEUSAGE, ZPROCESS)
15	com.apple.datausage.maps	Wifi In:0Wifi Out:0Wan In:95516Wan Out:48216	2021-06-06 14:16	DataUsage.sqlite : 0xC0C8B (Table: ZLIVEUSAGE, ZPROCESS)
16	com.apple.datausage.diagnostics	Wifi In:0Wifi Out:0Wan In:18027Wan Out:4214	2021-06-06 14:16	DataUsage.sqlite : 0xC0DEA (Table: ZLIVEUSAGE, ZPROCESS)
17	com.apple.datausage.location	Wifi In:0Wifi Out:0Wan In:2120938Wan Out:93136	2021-06-06 14:16	DataUsage.sqlite : 0xC107B (Table: ZLIVEUSAGE, ZPROCESS)
18	com.apple.datausage.appleid	Wifi In:0Wifi Out:0Wan In:801712Wan Out:51572	2021-06-06 10:00	DataUsage.sqlite : 0xC0CAD (Table: ZLIVEUSAGE, ZPROCESS)
19	com.apple.datausage.applepushservice	Wifi In:0Wifi Out:0Wan In:53446Wan Out:90992	2021-06-06 8:41	DataUsage.sqlite : 0xC0BEB (Table: ZLIVEUSAGE, ZPROCESS)
20	com.apple.datausage.dns	Wifi In:0Wifi Out:0Wan In:16566Wan Out:11228	2021-06-06 4:13	DataUsage.sqlite : 0xC0C0D (Table: ZLIVEUSAGE, ZPROCESS)
21	com.apple.datausage.siri	Wifi In:0Wifi Out:0Wan In:8557Wan Out:115810	2021-06-06 0:00	DataUsage.sqlite : 0xC0A8F (Table: ZLIVEUSAGE, ZPROCESS)
22	FaceTime	Wifi In:0Wifi Out:0Wan In:11733267Wan Out:16922028	2021-06-05 16:53	DataUsage.sqlite : 0xC0AB0 (Table: ZLIVEUSAGE, ZPROCESS)
23	FaceTime	Wifi In:0Wifi Out:0Wan In:924Wan Out:924	2021-06-05 16:44	DataUsage.sqlite : 0xC0AD4 (Table: ZLIVEUSAGE, ZPROCESS)
24	Safari	Wifi In:0Wifi Out:0Wan In:3619044Wan Out:366003	2021-06-05 7:49	DataUsage.sqlite : 0xC0AF4 (Table: ZLIVEUSAGE, ZPROCESS)
25	com.apple.datausage.iad	Wifi In:0Wifi Out:0Wan In:388Wan Out:300	2021-06-05 7:49	DataUsage.sqlite : 0xC0B56 (Table: ZLIVEUSAGE, ZPROCESS)
26	Safari	Wifi In:0Wifi Out:0Wan In:1033Wan Out:1544	2021-06-05 7:49	DataUsage.sqlite : 0xC0B16 (Table: ZLIVEUSAGE, ZPROCESS)
27	com.apple.datausage.location	Wifi In:0Wifi Out:0Wan In:6678Wan Out:3298	2021-06-05 7:49	DataUsage.sqlite : 0xC0B36 (Table: ZLIVEUSAGE, ZPROCESS)
28	com.apple.datausage.maps	Wifi In:0Wifi Out:0Wan In:132Wan Out:264	2021-06-05 1:16	DataUsage.sqlite : 0xC0B75 (Table: ZLIVEUSAGE, ZPROCESS)
29	com.apple.datausage.siri	Wifi In:0Wifi Out:0Wan In:615970Wan Out:1579595	2021-06-04 23:56	DataUsage.sqlite : 0xC1C90 (Table: ZLIVEUSAGE, ZPROCESS)
30	com.apple.datausage.dns	Wifi In:0Wifi Out:0Wan In:368195Wan Out:321436	2021-06-04 23:38	DataUsage.sqlite : 0xC2166 (Table: ZLIVEUSAGE, ZPROCESS)
31	com.apple.datausage.appleid	Wifi In:0Wifi Out:0Wan In:12473512Wan Out:5332674	2021-06-04 21:02	DataUsage.sqlite : 0xC2123 (Table: ZLIVEUSAGE, ZPROCESS)
32	com.apple.datausage.applepushservice	Wifi In:0Wifi Out:0Wan In:2216089Wan Out:3857470	2021-06-04 19:21	DataUsage.sqlite : 0xC1C4E (Table: ZLIVEUSAGE, ZPROCESS)
33	com.apple.datausage.general	Wifi In:0Wifi Out:0Wan In:92287Wan Out:36916	2021-06-04 11:23	DataUsage.sqlite : 0xC1C6F (Table: ZLIVEUSAGE, ZPROCESS)
34	com.apple.datausage.bluetooth	Wifi In:0Wifi Out:0Wan In:78568Wan Out:35444	2021-06-04 11:23	DataUsage.sqlite : 0xC2145 (Table: ZLIVEUSAGE, ZPROCESS)
35	com.apple.datausage.location	Wifi In:0Wifi Out:0Wan In:11811357Wan Out:832164	2021-06-04 1:19	DataUsage.sqlite : 0xC13E7 (Table: ZLIVEUSAGE, ZPROCESS)
36	com.apple.datausage.iad	Wifi In:0Wifi Out:0Wan In:138502Wan Out:50771	2021-06-04 1:18	DataUsage.sqlite : 0xC18C0 (Table: ZLIVEUSAGE, ZPROCESS)
37	FaceTime	Wifi In:0Wifi Out:0Wan In:51230159Wan Out:67427556	2021-06-03 18:01	DataUsage.sqlite : 0xC2209 (Table: ZLIVEUSAGE, ZPROCESS)
38	FaceTime	Wifi In:0Wifi Out:0Wan In:5940Wan Out:5940	2021-06-03 17:59	DataUsage.sqlite : 0xC222C (Table: ZLIVEUSAGE, ZPROCESS)
39	com.apple.datausage.media	Wifi In:0Wifi Out:0Wan In:3564Wan Out:9152	2021-06-03 17:59	DataUsage.sqlite : 0xC1D14 (Table: ZLIVEUSAGE, ZPROCESS)
40	com.apple.datausage.icloud	Wifi In:0Wifi Out:0Wan In:37504Wan Out:20146	2021-06-03 7:56	DataUsage.sqlite : 0xC21A6 (Table: ZLIVEUSAGE, ZPROCESS)
41	com.apple.datausage.location	Wifi In:0Wifi Out:0Wan In:10682Wan Out:8480	2021-06-03 7:46	DataUsage.sqlite : 0xC1A07 (Table: ZLIVEUSAGE, ZPROCESS)
42	com.apple.datausage.maps	Wifi In:0Wifi Out:0Wan In:1116635Wan Out:452533	2021-06-03 7:17	DataUsage.sqlite : 0xC1C2D (Table: ZLIVEUSAGE, ZPROCESS)
43	Weather	Wifi In:0Wifi Out:0Wan In:41871Wan Out:17159	2021-06-02 18:19	DataUsage.sqlite : 0xC1CB1 (Table: ZLIVEUSAGE, ZPROCESS)
44	Maps	Wifi In:0Wifi Out:0Wan In:136669Wan Out:36572	2021-06-02 18:19	DataUsage.sqlite : 0xC13C6 (Table: ZLIVEUSAGE, ZPROCESS)
45	com.apple.datausage.docsandsync	Wifi In:0Wifi Out:0Wan In:208051Wan Out:51009	2021-06-02 18:19	DataUsage.sqlite : 0xC1409 (Table: ZLIVEUSAGE, ZPROCESS)
46	com.apple.datausage.itunesmedia	Wifi In:0Wifi Out:0Wan In:272402Wan Out:37604	2021-06-02 18:19	DataUsage.sqlite : 0xC1A26 (Table: ZLIVEUSAGE, ZPROCESS)
47	Weather	Wifi In:0Wifi Out:0Wan In:217322Wan Out:76525	2021-06-02 18:19	DataUsage.sqlite : 0xC1B4B (Table: ZLIVEUSAGE, ZPROCESS)
48	com.apple.datausage.itunesmedia	Wifi In:0Wifi Out:0Wan In:1194168Wan Out:207249	2021-06-02 18:19	DataUsage.sqlite : 0xC19C7 (Table: ZLIVEUSAGE, ZPROCESS)
49	Maps	Wifi In:0Wifi Out:0Wan In:7435Wan Out:7222	2021-06-02 18:19	DataUsage.sqlite : 0xC19E8 (Table: ZLIVEUSAGE, ZPROCESS)
50	News	Wifi In:0Wifi Out:0Wan In:3074815Wan Out:250262	2021-06-02 18:19	DataUsage.sqlite : 0xC1AAA (Table: ZLIVEUSAGE, ZPROCESS)
51	com.apple.datausage.softwareupdate	Wifi In:0Wifi Out:0Wan In:132576Wan Out:40396	2021-06-02 18:19	DataUsage.sqlite : 0xC1D33 (Table: ZLIVEUSAGE, ZPROCESS)
52	Weather	Wifi In:0Wifi Out:0Wan In:15794Wan Out:12758	2021-06-02 18:19	DataUsage.sqlite : 0xC19A8 (Table: ZLIVEUSAGE, ZPROCESS)
53	News	Wifi In:0Wifi Out:0Wan In:22374Wan Out:19917	2021-06-02 18:19	DataUsage.sqlite : 0xC1ACB (Table: ZLIVEUSAGE, ZPROCESS)
54	com.apple.AppStore	Wifi In:0Wifi Out:0Wan In:83942Wan Out:46392	2021-06-02 18:18	DataUsage.sqlite : 0xC189F (Table: ZLIVEUSAGE, ZPROCESS)
55	com.apple.datausage.appleid	Wifi In:0Wifi Out:0Wan In:464718Wan Out:144279	2021-06-02 18:18	DataUsage.sqlite : 0xC1A47 (Table: ZLIVEUSAGE, ZPROCESS)
56	News	Wifi In:0Wifi Out:0Wan In:96139Wan Out:31671	2021-06-02 18:18	DataUsage.sqlite : 0xC1803 (Table: ZLIVEUSAGE, ZPROCESS)
57	com.apple.datausage.media	Wifi In:0Wifi Out:0Wan In:61728Wan Out:17713	2021-06-02 18:18	DataUsage.sqlite : 0xC1A8A (Table: ZLIVEUSAGE, ZPROCESS)
58	com.apple.datausage.findmyiphone	Wifi In:0Wifi Out:0Wan In:192172Wan Out:188166	2021-06-02 18:18	DataUsage.sqlite : 0xC18E1 (Table: ZLIVEUSAGE, ZPROCESS)
59	com.apple.datausage.itunesmedia	Wifi In:0Wifi Out:0Wan In:126574Wan Out:47606	2021-06-02 18:18	DataUsage.sqlite : 0xC1902 (Table: ZLIVEUSAGE, ZPROCESS)
60	com.apple.datausage.general	Wifi In:0Wifi Out:0Wan In:268257Wan Out:105736	2021-06-02 18:18	DataUsage.sqlite : 0xC1986 (Table: ZLIVEUSAGE, ZPROCESS)
61	com.apple.AppStore	Wifi In:0Wifi Out:0Wan In:5785Wan Out:7221	2021-06-02 18:18	DataUsage.sqlite : 0xC1967 (Table: ZLIVEUSAGE, ZPROCESS)
62	com.apple.datausage.messages	Wifi In:0Wifi Out:0Wan In:486440Wan Out:206677	2021-06-02 18:18	DataUsage.sqlite : 0xC1AEA (Table: ZLIVEUSAGE, ZPROCESS)
63	Safari	Wifi In:0Wifi Out:0Wan In:21211298Wan Out:1028653	2021-06-02 11:55	DataUsage.sqlite : 0xC1CD1 (Table: ZLIVEUSAGE, ZPROCESS)
64	Safari	Wifi In:0Wifi Out:0Wan In:11571Wan Out:22604	2021-06-02 11:55	DataUsage.sqlite : 0xC1CF3 (Table: ZLIVEUSAGE, ZPROCESS)
65	Weather	Wifi In:0Wifi Out:0Wan In:22180Wan Out:5864	2021-05-29 17:19	DataUsage.sqlite : 0xC1865 (Table: ZLIVEUSAGE, ZPROCESS)
66	com.apple.datausage.appleid	Wifi In:0Wifi Out:0Wan In:57997Wan Out:10813	2021-05-28 20:16	DataUsage.sqlite : 0xC1B2B (Table: ZLIVEUSAGE, ZPROCESS)
67	com.apple.datausage.telephony	Wifi In:0Wifi Out:0Wan In:2283350Wan Out:51931203	2021-05-28 8:36	DataUsage.sqlite : 0xC1A68 (Table: ZLIVEUSAGE, ZPROCESS)
83	Maps	Wifi In:0Wifi Out:0Wan In:234248Wan Out:18996	2021-05-22 13:41	DataUsage.sqlite : 0xC1384 (Table: ZLIVEUSAGE, ZPROCESS)
84	com.apple.datausage.media	Wifi In:0Wifi Out:0Wan In:37971Wan Out:11188	2021-05-22 13:41	DataUsage.sqlite : 0xC1947 (Table: ZLIVEUSAGE, ZPROCESS)
85	com.apple.datausage.security	Wifi In:0Wifi Out:0Wan In:19906Wan Out:5675	2021-05-22 13:41	DataUsage.sqlite : 0xC148C (Table: ZLIVEUSAGE, ZPROCESS)
88	com.apple.datausage.siri	Wifi In:0Wifi Out:0Wan In:64179Wan Out:15754	2021-05-18 6:08	DataUsage.sqlite : 0xC1345 (Table: ZLIVEUSAGE, ZPROCESS)

Fig. 6.46 iPhone X listing of 71 removed log entries from 72 h extraction

#	Application	Body	Time (UTC-3)	Source file information
1	com.apple.datausage.dns	Wifi In:0 Wifi Out:0 Wan In:31188 Wan Out:25417	11-Jun-21 8:52:35 AM	DataUsage.sqlite-wal : 0x225EEC (Table: ZLIVEUSAGE, ZPROCESS, Size: 2253672 bytes)

Fig. 6.47 iPhone X listing of 1 new log entry from the 7 days extraction

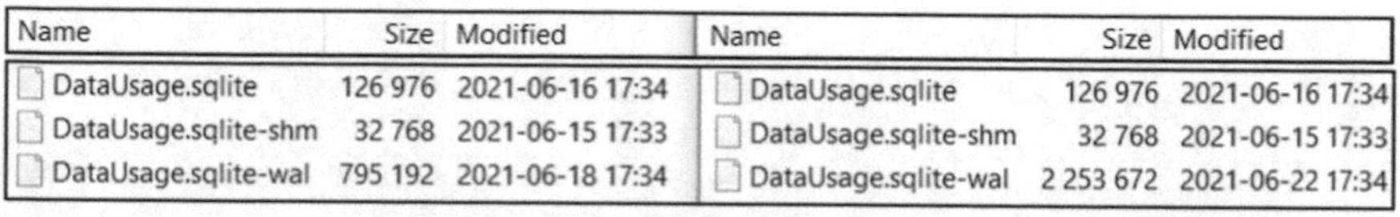

Name	Size	Modified	Name	Size	Modified
DataUsage.sqlite	126 976	2021-06-16 17:34	DataUsage.sqlite	126 976	2021-06-16 17:34
DataUsage.sqlite-shm	32 768	2021-06-15 17:33	DataUsage.sqlite-shm	32 768	2021-06-15 17:33
DataUsage.sqlite-wal	795 192	2021-06-18 17:34	DataUsage.sqlite-wal	2 253 672	2021-06-22 17:34

Fig. 6.48 iPhone X comparison of DataUsage.sqlite database files 72 h versus 7 days

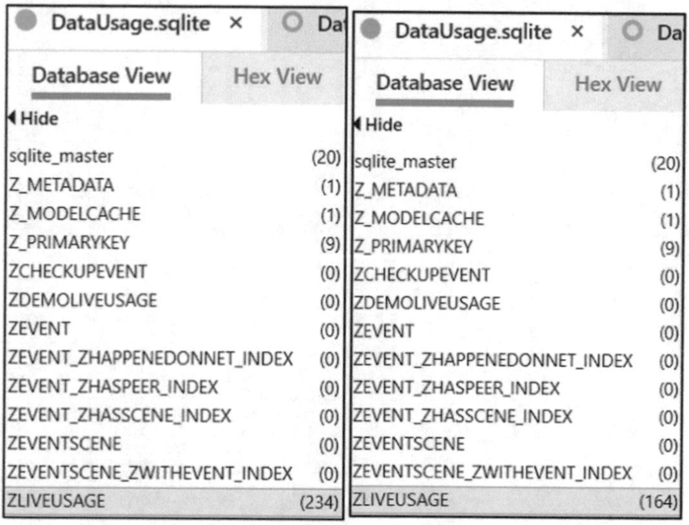

Fig. 6.49 iPhone X ZLIVEUSAGE entries from the DataUsage.sqlite database 72 h versus 7 days

In comparing the locations where these 5 thumbnails images were stored (2 folder paths and 3 data files), each had a common modified time of 2021-06-19 05:37, which was noted in the 7 days extraction. In addition, only the 100APPLE folders saw a decrease in size of 5126 and 43,566 B, while the '.ithmb' files sizes remained the same. A comparison of these folders and data files from the iPhone X's 72 h and 7 day extractions are shown in Fig. 6.54.

The remaining 61 image files were associated to a common source path of /private/var/mobile/ Library within the iPhone X's file system. In reviewing this source path, the 61 images divided between three different locations, all of which were found to be associated to the native iOS Messages application. Of these 61 images 4 had an associated source path of /private/var/mobile/Library/SMS/Attachments/, shown in Fig. 6.55.

In analyzing this location within the iPhone X's file system, the removal of all four image files was confirmed. In addition, each folder that contained one of the removed images not only saw a reduction in folder size but also had a new modified time of 2021-06-20 03:00, displayed in Fig. 6.56.

Z_PK	ZHASPROCESS	ZTIMESTAMP	ZWWANIN	ZWWANOUT	Z_PK	ZHASPROCESS	ZTIMESTAMP	ZWWANIN	ZWWANOUT
873	96	2021-06-04 4:18	138502	50771	943	88	2021-06-08 14:01	337502	149881
872	80	2021-06-05 0:02	12473512	5332674	942	83	2021-06-09 0:07	80286	19370
871	73	2021-06-04 14:23	78568	35444	941	82	2021-06-08 7:58	66721	34583
870	51	2021-06-04 4:19	11811357	832164	940	81	2021-06-08 14:57	1251804	220268
869	43	2021-06-04 22:21	2216089	3857470	939	80	2021-06-09 0:05	13792257	8554519
868	39	2021-06-04 14:23	92287	36916	937	75	2021-06-09 0:07	24836590	1398576
867	38	2021-06-05 2:56	615970	1579595	936	74	2021-06-09 0:07	12806	24543
866	36	2021-06-05 2:38	368195	321436	935	73	2021-06-08 7:58	82126	37380
865	86	2021-05-18 9:12	14605	8105	934	70	2021-06-08 23:11	65091364	86645768
864	82	2021-06-03 10:56	37504	20146	933	69	2021-06-08 23:11	7304	7304
861	70	2021-06-03 21:01	51230159	67427556	932	54	2021-06-08 7:58	25105	23229
860	69	2021-06-03 20:59	5940	5940	931	51	2021-06-09 1:12	14204177	937788
859	68	2021-06-03 20:59	3564	9152	930	49	2021-06-09 1:13	1220565	505885
856	52	2021-06-03 10:46	98138	49612	929	43	2021-06-08 23:24	2399719	4127747
855	49	2021-06-03 10:17	1116635	452533	928	40	2021-06-09 0:07	255649	63871
853	42	2021-06-03 10:46	10682	8480	927	39	2021-06-08 7:58	96226	38765
849	108	2021-06-02 21:19	41871	17159	926	38	2021-06-08 3:48	631916	1732575
848	107	2021-05-28 23:16	57997	10813	925	36	2021-06-09 1:12	388490	338986
847	105	2021-06-02 21:18	61728	17713	921	67	2021-06-08 0:23	2311234	52537875
846	103	2021-06-02 21:19	217322	76525	917	108	2021-06-06 17:23	49911	20721
845	101	2021-06-02 21:19	15794	12758	916	107	2021-06-06 17:16	71109	14656
844	100	2021-06-02 21:19	272402	37604	914	105	2021-06-06 17:16	78055	21548
843	99	2021-06-02 21:18	464718	144279	913	103	2021-06-06 17:23	239521	84372
842	98	2021-06-02 21:18	83942	46392	912	102	2021-06-06 17:23	29695	7892
841	97	2021-06-02 21:18	96139	31671	911	101	2021-06-06 17:23	17559	14584
839	94	2021-06-02 21:18	192172	188166	910	100	2021-06-06 17:17	320726	46014
838	93	2021-06-02 21:18	126574	47606	909	99	2021-06-06 17:16	588556	181845
837	91	2021-05-22 16:41	19906	5675	908	98	2021-06-06 17:17	115237	66251
836	90	2021-06-02 21:18	5785	7221	907	97	2021-06-06 17:21	109867	36228
835	88	2021-06-02 21:18	268257	105736	906	96	2021-06-06 17:21	139118	51299
834	83	2021-05-18 9:08	64179	15754	905	95	2021-06-06 17:16	350040	30144
833	81	2021-06-02 21:19	1194168	207249	904	94	2021-06-06 17:16	232738	229933
832	78	2021-06-02 21:19	136669	36572	903	93	2021-06-06 17:16	175742	63646
831	77	2021-06-02 21:19	7435	7222	902	92	2021-06-06 17:16	66972	19072
830	75	2021-06-02 14:55	21211298	1028653	901	91	2021-06-06 17:16	23814	6927
829	74	2021-06-02 14:55	11571	22604	900	90	2021-06-06 17:16	7321	8874
827	55	2021-06-02 21:19	3074815	250262	895	78	2021-06-06 17:23	144484	40113
826	54	2021-06-02 21:19	22374	19917	894	77	2021-06-06 17:16	8093	8652
824	50	2021-06-02 21:18	486440	206677	892	68	2021-06-06 17:17	4356	11396
823	48	2021-06-02 21:19	132576	40396	891	55	2021-06-06 17:23	3124369	258962
821	40	2021-06-02 21:19	208051	51009	888	50	2021-06-06 17:16	511643	219288
817	104	2021-05-30 15:49	1678217	165760	886	48	2021-06-06 17:17	158559	50567
815	102	2021-05-29 20:19	22180	5864	874	42	2021-06-05 4:16	11271	8876
808	95	2021-05-22 16:41	234248	18996	856	52	2021-06-03 10:46	98138	49612
776	92	2021-06-11 23:59	140	16335	817	104	2021-05-30 15:49	1678217	165760
775	106	2021-06-11 23:59	140	16335	719	43	2021-06-11 7:57	526631	10900196
774	78	2021-06-11 23:59	140	16335	720	38	2021-06-11 10:18	407	42135
773	115	2021-06-11 23:59	140	16335	721	36	2021-06-11 11:52	31188	25417
772	77	2021-06-11 23:58	658	884	722	70	2021-06-11 11:52	0	100516
771	36	2021-06-11 23:58	809	263	751	69	2021-06-11 17:58	0	5368
770	77	2021-06-11 23:58	660	270	752	42	2021-06-11 19:25	228	456
769	75	2021-06-11 21:23	280	32670	753	48	2021-06-11 19:25	1424	33714
768	74	2021-06-11 21:22	202	316	754	101	2021-06-11 19:25	2075	1874
767	99	2021-06-11 19:28	280	32670	755	49	2021-06-11 19:25	3220	71539
766	51	2021-06-11 19:27	2691	115636	756	55	2021-06-11 19:25	280	17272
765	40	2021-06-11 19:26	10488	35638	757	103	2021-06-11 19:25	420	20661
764	100	2021-06-11 19:26	280	32670	758	68	2021-06-11 19:25	322	1242
763	94	2021-06-11 19:26	280	32670	759	50	2021-06-11 19:25	1456	32267
762	88	2021-06-11 19:26	7368	40222	760	81	2021-06-11 19:26	1916	83177
761	93	2021-06-11 19:26	140	16335	761	93	2021-06-11 19:26	140	16335
760	81	2021-06-11 19:26	1916	83177	762	88	2021-06-11 19:26	7368	40222
759	50	2021-06-11 19:25	1456	32267	763	94	2021-06-11 19:26	280	32670
758	68	2021-06-11 19:25	322	1242	764	100	2021-06-11 19:26	280	32670

Fig. 6.50 iPhone X ZLIVEPROCESS table comparison showing 71 lost entries 72 h versus 7 days

Z_PK	ZHASPROCESS	ZTIMESTAMP	ZWWANIN	ZWWANOUT	Z_PK	ZHASPROCESS	ZTIMESTAMP	ZWWANIN	ZWWANOUT
719	43	2021-06-11 7:57	526631	10900196	719	43	2021-06-11 7:57	526631	10900196
720	38	2021-06-11 10:18	407	42135	720	38	2021-06-11 10:18	407	42135
721	36	2021-06-11 11:52	31188	25417	721	36	2021-06-11 11:52	31188	25417
722	70	2021-06-11 11:52	0	100516	722	70	2021-06-11 11:52	0	100516
751	69	2021-06-11 17:58	0	5368	751	69	2021-06-11 17:58	0	5368
752	42	2021-06-11 19:25	228	456	752	42	2021-06-11 19:25	228	456
753	48	2021-06-11 19:25	1424	33714	753	48	2021-06-11 19:25	1424	33714
754	101	2021-06-11 19:25	2075	1874					
755	49	2021-06-11 19:25	3220	71539					

Fig. 6.51 iPhone X ZLIVEPROCESS table comparison showing 1 new entry 72 h versus 7 days

Source: /private/var/mobile/Media/PhotoData/Metadata/DCIM/100APPLE/				
#	Size (bytes)	MD5	Created-Date	Path and/or Name
37	5126	5bcb7789007b90677c809ae17ae13848	19-05-21	IMG_0094.THM
Source: /private/var/mobile/Media/PhotoData/Thumbnails/				
#	Size (bytes)	MD5	Created-Date	Path and/or Name
66	3186	c2b27559135a422508659a1333e0a04e		3306.ithmb/thumb_93.bmp
64	31566	3e7e88fa590b634a054bb3beb5590f28		3314.ithmb/thumb_93.bmp
65	28866	6681ceaeccfc8a5378a3f4f6db14b29c		4031.ithmb/thumb_93.bmp
6	43566	690abd89f857eb5f71de48e22005df70	19-05-21	/V2/DCIM/100APPLE/IMG_0094.MOV/5005.JPG

Fig. 6.52 iPhone X 5 lost thumbnails from 72 h extraction

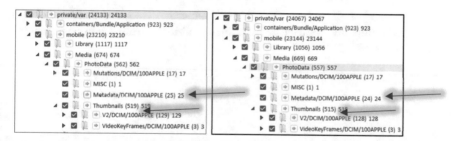

Fig. 6.53 iPhone X 5 thumbnail file system locations 72 h versus 7 days

Name	Size	Modified	Name	Size	Modified
100APPLE	1 728 507	2021-06-18 05:37	100APPLE	1 723 381	2021-06-19 05:37
100APPLE	6 505 036	2021-06-18 05:37	100APPLE	6 461 470	2021-06-19 05:37
3306.ithmb	604 416	2021-06-18 05:37	3306.ithmb	604 416	2021-06-19 05:37
3314.ithmb	6 149 376	2021-06-18 05:37	3314.ithmb	6 149 376	2021-06-19 05:37
4031.ithmb	5 534 976	2021-06-18 05:37	4031.ithmb	5 534 976	2021-06-19 05:37

Fig. 6.54 iPhone X lost thumbnail locations 72 h versus 7 days

	Source: /private/var/mobile/Library/SMS/Attachments/			
#	Size (bytes)	MD5	Created Date	Path and/or Name
13	1344402	9a086d5ec2225d69f106a75701b54af6	06-06-21	62/02/CC68ABEE-6D1B-4439-96CA-01A64B4FF3D0/CC68ABEE-6D1B-4439-96CA-01A64B4FF3D0.pvt/64469311993__7813038D-F1EC-4A37-859D-43955081F49B.HEIC
19	1992095	9820971fdb3853cba771e72b4d8ef2f9	09-06-21	75/05/72C91C3C-DA84-48FA-8E46-890E804FF687/72C91C3C-DA84-48FA-8E46-890E804FF687.pvt/64496376973__C6FADAEA-E498-475F-9D0D-659FE0F692A0.HEIC
44	1517376	3aa6bf83fb23d53ef542d3bf23efaf68	06-06-21	97/07/0152D003-E70E-4EDB-9F1E-4DF0303FA8DD/0152D003-E70E-4EDB-9F1E-4DF0303FA8DD.pvt/IMG_0170.HEIC
17	547846	2db1c02512029ffba9723c6c13628ce8	07-06-21	9b/11/E1427D62-D304-484F-B694-48E64631CF88/E1427D62-D304-484F-B694-48E64631CF88.pvt/64478969545__0A655073-6CE2-4238-9644-20F00FAA00E5.HEIC

Fig. 6.55 iPhone X 4 lost image files from 72 h extraction

Name	Size	Modified	Name	Size	Modified
CC68ABEE-6D1B-4439-96CA-01A64B4FF3D0	7 498 002	2021-06-06 14:25	CC68ABEE-6D1B-4439-96CA-01A64B4FF3D0	3 748 873	2021-06-20 03:00
72C91C3C-DA84-48FA-8E46-890E804FF687	7 860 036	2021-06-09 17:36	72C91C3C-DA84-48FA-8E46-890E804FF687	3 929 890	2021-06-20 03:00
0152D003-E70E-4EDB-9F1E-4DF0303FA8DD	7 510 300	2021-06-06 12:23	0152D003-E70E-4EDB-9F1E-4DF0303FA8DD	3 755 022	2021-06-20 03:00
E1427D62-D304-484F-B694-48E64631CF88	4 258 988	2021-06-07 17:15	E1427D62-D304-484F-B694-48E64631CF88	2 129 366	2021-06-20 03:00

Fig. 6.56 iPhone X /private/var/mobile/Library/SMS/Attachments/ sub-subfolders 72 h versus 7 days

The second of the three locations contained five image files located in /private/var/mobile/Library/Caches/com.apple.MobileSMS/BrowserSnapshots/, shown in Fig. 6.57.

Source: /private/var/mobile/Library/Caches/com.apple.MobileSMS/BrowserSnapshots/				
#	Size (bytes)	MD5	Created Date	Name
1	12716	77c56a3e354c5160f76bb8ca5a0237f2	11-06-21	10465042522214366315.png
2	302	7c13a3b780a7e0dedf12ebb5c1de4c20	06-06-21	14019782735580945992.png
3	21592	2ffab18703a54d5cbb7b4c7bf09d4602	07-06-21	15799904270332018302.png
4	302	7c13a3b780a7e0dedf12ebb5c1de4c20	06-06-21	16242543410136845744.png
5	25173	538d625089b0c13cf5d2e04db23f7e74	08-06-21	30460888966689778919.png

Fig. 6.57 iPhone X 5 lost image files from 72 h extraction

Name	Size	Modified	Name	Size	Modified
BrowserSnapshots	60 085	2021-06-11 15:27	Plugins	0	2021-05-12 16:14
Plugins	0	2021-05-12 16:14	Previews	71 573 756	2021-05-12 21:03
Previews	98 452 779	2021-05-12 21:03			

Fig. 6.58 iPhone X removal of BrowserSnapshots subfolder 72 h versus 7 days

Name	Size	Modified	Name	Size	Modified
Attachments	63 136 887	2021-06-11 15:27	Attachments	36 257 864	2021-06-11 15:27

Fig. 6.59 iPhone X /private/var/mobile/Library/Caches/com.apple.MobileSMS//Previews/Attachments/comparison 72 h versus 7 days

In reviewing this location in both the 72 h and 7 day extractions, it was found that the /private/var/mobile/Library/ Caches/com.apple.MobileSMS/BrowserSnapshots/ subfolder had been completely removed thus confirming the loss of these 5 images, shown in Fig. 6.58.

The third location contained 52 image files located in /private/var/mobile/Library/Caches/ com.apple.MobileSMS/Previews/Attachments/. Of these 52 images, 34 had a file extension of '.ktx'. As previously stated in this research paper, '.ktx' files are iOS snapshots, however; in this context, the '.ktx' files appear to represent a thumbnail image.

In comparing the overall size difference of the Attachments folder from the iPhone X's 72 h and 7 day extractions, although there was no change in modified time, the folder's size did decrease by 26,879,023 B, shown in Fig. 6.59.

In comparing the iPhone X's 72 h and 7 day extractions, a reduction in size of 26,939,108 B and a new modified time of 2021-06-20 03:00 was observed in the 7 day extraction for the /private/var/mobile/Library/Caches/ com.apple.MobileSMS/ folder, shown in Fig. 6.60.

In reviewing each of the 52 images folder locations, their removal was confirmed, shown in Fig. 6.61.

Videos: The artifacts of interest comparison between the iPhone X's 72 h and 7 day extractions identified a decrease in 10 videos. In comparing the video files from both extractions, the 10 videos were identified, tagged and reviewed. The 10 video files were associated to three different source path locations within the iPhone X's file system. A listing of these 10 video can be seen in Fig. 6.62.

Each of the three file system locations were reviewed, and the removal of the 10 videos was confirmed, shown in Fig. 6.63.

Name	Size	Modified	Name	Size	Modified
com.apple.MobileSMS	98 512 864	2021-06-06 12:23	com.apple.MobileSMS	71 573 756	2021-06-20 03:00

Fig. 6.60 iPhone X com.apple.MobileSMS folder comparison 72 h versus 7 days

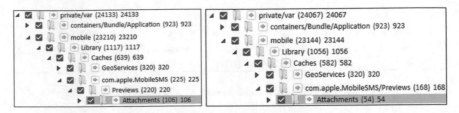

Fig. 6.61 iPhone X 52 removed images comparison from /com.apple.MobileSMS/Previews /Attachments/ 72 h versus 7 days

#	Name	Size (bytes)	Path and/or Name	MD5 Hash Value	Created-Time (UTC-3)	Deleted
	Source: /private/var/mobile/Media/DCIM/100APPLE/					
7	IMG_0094.MOV	950085799	IMG_0094.MOV	5307971301cd05f9ebf42c053b7fd0e5	19-05-21 19:23	Trash
	Source: /private/var/mobile/Library/SMS/Attachments/					
5	64478969545__0A655073-6CE2-4238-9644-20F00FAA00E5_1.MOV	1581520	9b/11/E1427D62-D304-484F-B694-48E64631CF88/E1427D62-D304-484F-B694-48E64631CF88.pvt/64478969545__0A655073-6CE2-4238-9644-20F00FAA00E5.MOV	83cb98506d6a556564381c4ff7930107	07-06-21 17:14	
#	IMG_0170_1.MOV	2237646	97/07/0152D003-E70E-4EDB-9F1E-4DF0303FA8DD/0152D003-E70E-4EDB-9F1E-4DF0303FA8DD.pvt/IMG_0170.MOV	acc1426869ccd1435576c098afe86cb9	06-06-21 12:23	
6	64496376973__C6FADAEA-E498-475F-9D0D-659FE0F692A0.MOV	1937795	75/05/72C91C3C-DA84-48FA-8E46-890E804FF687/72C91C3C-DA84-48FA-8E46-890E804FF687.pvt/64496376973__C6FADAEA-E498-475F-9D0D-659FE0F692A0.MOV	5eaa453c89e5bad83f8ba7e5ec8bb51b	09-06-21 17:36	
3	64469311993__7813038D-F1EC-4A37-859D-43955081F49B_1.MOV	2404471	62/02/CC68ABEE-6D1B-4439-96CA-01A64B4FF3D0/CC68ABEE-6D1B-4439-96CA-01A64B4FF3D0.pvt/64469311993__7813038D-F1EC-4A37-859D-43955081F49B.MOV	1e378241273e35399cd578812169cd7f	06-06-21 14:25	
	Source: /private/var/mobile/Library/Caches/com.apple.MobileSMS/Previews/Attachments/					
4	64478969545__0A655073-6CE2-4238-9644-20F00FAA00E5.MOV	1581520	9b/11/E1427D62-D304-484F-B694-48E64631CF88/64478969545__0A655073-6CE2-4238-9644-20F00FAA00E5-preview.pvt/64478969545__0A655073-6CE2-4238-9644-20F00FAA00E5.MOV	83cb98506d6a556564381c4ff7930107	07-06-21 17:14	
9	IMG_0170.MOV	2237646	97/07/0152D003-E70E-4EDB-9F1E-4DF0303FA8DD/IMG_0170-preview.pvt/IMG_0170.MOV	acc1426869ccd1435576c098afe86cb9	06-06-21 12:23	
8	IMG_0154.MOV	1410940	95/05/12713F69-1BEF-46FF-AEC9-2C5DB21B0D7D/IMG_0154-preview.pvt/IMG_0154.MOV	a57a3e4185e141a306e81c8747b10fda	01-06-21 14:57	
2	64469311993__7813038D-F1EC-4A37-859D-43955081F49B.MOV	2404471	62/02/CC68ABEE-6D1B-4439-96CA-01A64B4FF3D0/64469311993__7813038D-F1EC-4A37-859D-43955081F49B-preview.pvt/64469311993__7813038D-F1EC-4A37-859D-43955081F49B.MOV	1e378241273e35399cd578812169cd7f	06-06-21 14:25	
1	64392340500__A94A8FBD-6BFA-4B25-95DB-528EDA023A1B.MOV	1305637	30/00/F1AC60E2-7677-40D7-A9BE-8AB09C0FF855/64392340500__A94A8FBD-6BFA-4B25-95DB-528EDA023A1B-preview.pvt/64392340500__A94A8FBD-6BFA-4B25-95DB-528EDA023A1B.MOV	97d755be5e779389f7813c1368ca9201	28-05-21 16:36	

Fig. 6.62 iPhone X 10 removed video files 72 h extraction

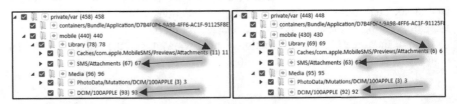

Fig. 6.63 iPhone X 10 removed video filse 72 h versus 7 days

Name	Size	Modified	Name	Size	Modified
100APPLE	2 367 599 057	2021-06-18 05:37	100APPLE	1 417 513 258	2021-06-19 05:37

Fig. 6.64 iPhone X /private/var/mobile/Media/DCIM/100APPLE/ comparison folder 72 h versus 7 days

In reviewing the first location, /private/var/mobile/Media/DCIM/100APPLE, a reduction in folder size as well as a new modified time of 2021-06-19 05:37 was noted in the 7 days extraction, shown in Fig. 6.64.

The other two folder locations that contained the nine remaining video files are the same folder locations that were reviewed for the lost image files. Changes in these folder sizes a well as modified times can be seen in Figs. 6.60 and 6.61.

Overall, in the case of iPhoneX analysis, the changes in the artifacts of interest were able to be confirmed through the application and database analysis conducted. In total, a decrease in 1108 application usage log entries, 165 log entries, 353 images and 67 videos was confirmed as well as one increase in log entries. The removal of only one log entry was not able to be confirmed as it had no associated source file. The verification of incremental auto vacuum mode on the knowledgeC.db, interactionC.db, and DataUsage.sqlite databases further supports the loss of log entries and noted decreases in database size.

6.5.2   Timeline and iOS Analysis

Once the comparison analysis and modification times of the iOS database files and paths were noted, the unified logs from each of the iPhone extractions were parsed using the forensic analysis tool Cellebrite Inspector. The findings of this analysis are presented below.

6.5.2.1   iPhone 7 Analysis

Vacuuming

The iOS unified logs from the iPhone 7's 7 days extraction were reviewed to verify the usage of SQLite vacuuming. In total, 1465 unified log messages contained the word vacuum, and of those 960 were specifically related to auto-vacuuming. A sample of these log entries are shown in Fig. 6.65.

The full message content of the second entry reads: _{"msg": "incrementalVacuum", "event": "elapsed", "begin_mach": "14251185579464", "end_mach": "14251185580475", "elapsed_s": "0.000042125", "pages": "3", "path": "/var/root/Library/Caches/locationd/gyroCal.db"} located in the log entry at /var/db/uuidtext/F6/E94E0F9F14383D948A6C21AC5730B4.

Type	Date & Time	Process	Message
	2017-10-22 13:28:49.500912-0300	apsd	2017-10-22 13:28:49 -0300 apsd[84]: APSMessageStore - Enabling auto vacuum.
	2017-10-22 13:38:44.484062-0300	locationd	{"msg":"incrementalVacuum", "result":"0"}
	2017-10-22 13:38:44.488072-0300	locationd	{"msg":"incrementalVacuum", "event":"elapsed", "begin_mach":"14261185679464", "end_mach":"14261...
	2017-10-22 13:38:49.565684-0300	apsd	2017-10-22 13:38:49 -0300 apsd[84]: APSMessageStore - Enabling auto vacuum.
	2017-10-22 13:48:49.596642-0300	apsd	2017-10-22 13:48:49 -0300 apsd[84]: APSMessageStore - Enabling auto vacuum.
	2017-10-22 13:58:49.628602-0300	apsd	2017-10-22 13:58:49 -0300 apsd[84]: APSMessageStore - Enabling auto vacuum.
	2017-10-22 14:08:49.678327-0300	apsd	2017-10-22 14:08:49 -0300 apsd[84]: APSMessageStore - Enabling auto vacuum.
	2017-10-22 14:18:49.705048-0300	apsd	2017-10-22 14:18:49 -0300 apsd[84]: APSMessageStore - Enabling auto vacuum.
	2017-10-22 14:28:49.769855-0300	apsd	2017-10-22 14:28:49 -0300 apsd[84]: APSMessageStore - Enabling auto vacuum.
	2017-10-22 14:38:49.808193-0300	apsd	2017-10-22 14:38:49 -0300 apsd[84]: APSMessageStore - Enabling auto vacuum.

Fig. 6.65 iPhone 7 auto-vacuum entries in unified logs

⊟ Search Hits

auto.vacuum: 0 hits

auto_vacuum: 43 hits (27 files)

autovacuum: 18 hits (9 files)

Totals: 61 hit (33 files)

Fig. 6.66 iPhone 7 search hit results auto-vacuum

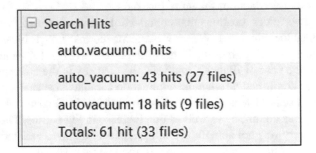

Fig. 6.67 iPhone 7 search hit results auto-vacuum

In conducting a deep search of the logs however, multiple instances of auto-vacuuming using PRAGMA states were located. With these statements not appearing specifically in the unified logs, it appears that this level of detail is excluded from the message content. The results of the search performed and examples of the PRAGMA statements are provided in Figs. 6.66 and 6.67.

com.apple.MobileSMS/Previews/Attachments

In conducting the timeline and unified logs analysis, the deletion of data from the /private/var/mobile/Library/Caches/com.apple.MobileSMS/Preview/Attachments file path was further confirmed with an entry in the unified logs with the same modified time. The entry was located in /private/var/db/uuidtext/61/4F2CCB08343 EDFB243C596ED4F9123 in the iPhone 7's 7 days extraction. The log entry is

%@kIMContactsContactIdentifierKeyWarningASSERTION FAILED: %@IMAutomaticHistoryDeletionAgent Launched!com.apple.notifyd.matchingv16@?0@"NSObject<OS_xpc_object>"8Cleansing orphaned attachmentsDeleting previews older than last 200Finished previews older than last 200Cleansing orphaned sticker transfer user infoIMAutomativeHistoryDeletionAgentCleansing orphaned sticker attachmentsCleansing browser snapshot cachedeleting stale ChatKit previewsdeleting stale ChatKit

Fig. 6.68 iPhone 7 /private/var/db/uuidtext/61/4F2CCB08343EDFB243C596ED4F9123 from 7 days extraction

	Date	Message	Process Name	Process Path	Message Type
	2021-06-22 13:04:01.16379...	APSMessageStore - Enabling auto vacuum.	apsd	/System/Library/PrivateFra...	Default
	2021-06-22 12:54:01.13323...	APSMessageStore - Enabling auto vacuum.	apsd	/System/Library/PrivateFra...	Default
	2021-06-22 12:44:01.13261...	APSMessageStore - Enabling auto vacuum.	apsd	/System/Library/PrivateFra...	Default

Fig. 6.69 iPhone X auto-vacuum entries in unified logs

for the IMAutomaticHistoryDeletionAgent, and in the log entry, the deletion of previews and snapshots is executed. A portion of the log file can be seen in Fig. 6.68.

6.5.2.2 iPhone X

Vacuuming

The iOS unified logs from the iPhone X's 7 days extraction were reviewed to verify the usage of SQLite vacuuming. In total, 3791 unified log messages contained the word vacuum, and of those 76 were specifically related to auto-vacuuming. A sample of the 76 log entries are shown in Fig. 6.69.

The full message content of the first entry reads: _xpc_activity_set_state: send new state to CTS: com.apple.message.db.vacuum (107e55550), 5, which was found in the log /private/var/db/diagnostics/Persist/0000000000000367.tracev3.

In conducting a deep search of the logs however, multiple instances of auto-vacuuming using PRAGMA states were located. With these statements not appearing specifically in the unified logs, it appears that this level of detail is excluded from the message content. The results of the search performed and examples of the PRAGMA statements are provided in Figs. 6.70 and 6.71.

knowledgeC.db, interactionC.db, and DataUsage.sqlite

While the unified logs do not provide a clear reason for the modified times of the knowledgeC.db, interactionC.db and DataUsage.sqlite database files, what can be correlated, is that the modification time of each of the database files occurred during one of iPhone X's timed extractions, demonstrated in Table 6.5.

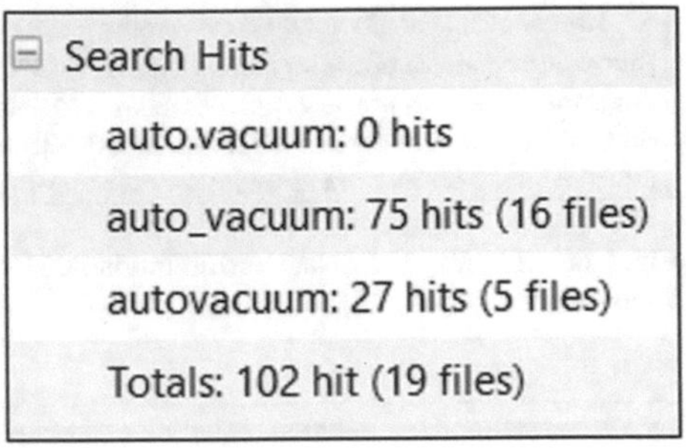

Fig. 6.70 iPhone X search hit results auto-vacuum

Search Hits

auto.vacuum: 0 hits

auto_vacuum: 75 hits (16 files)

autovacuum: 27 hits (5 files)

Totals: 102 hit (19 files)

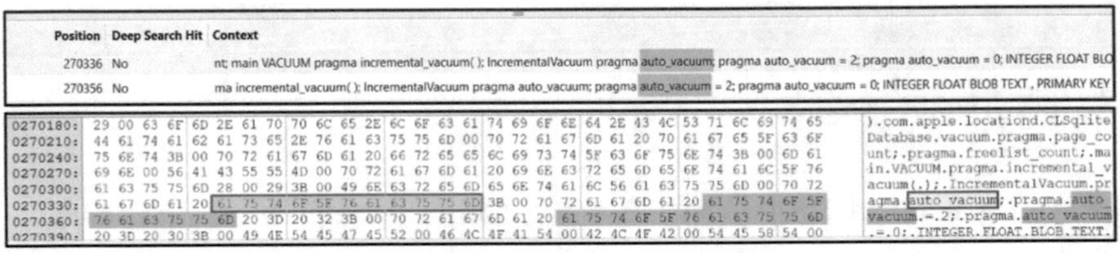

Fig. 6.71 iPhone X search hit results auto-vacuum 2

Table 6.5 iPhone X comparison of database file modification times with extraction times

Extraction timeframe	Database files or file path	Modification times in ADT	Extraction time in ADT
24 h	knowledgeC.db and knowledgeC.db-wal	2021-06-16 17:46	2021-06-16 17:34-18:05
24 h	interactionC.db-shm and interactionC.db-wal	2021-06-16 17:39	2021-06-16 17:34-18:05
24 h	DataUsage.sqlite and DataUsage.sqlite-wal	2021-06-16 17:34	2021-06-16 17:34-18:05
72 h	knowledgeC.db and knowledgeC.db-wal	2021-06-18 17:37 and 2021-06-18 17:46	2021-06-18 17:34-19:03
72 h	interactionC.db-shm and interactionC.db-wal	2021-06-18 17:39	2021-06-18 17:34-19:03
72 h	DataUsage.sqlite-wal	2021-06-18 17:34	2021-06-18 17:34-19:03
7 days	knowledgeC.db and knowledgeC.db-wal	2021-06-22 17:44 and 2021-06-22 17:46	2021-06-22 17:34-18:02
7 days	interactionC.db-shm and interactionC.db-wal	2021-06-22 17:39 and 2021-06-22 17:39	2021-06-22 17:34-18:02
7 days	DataUsage.sqlite-wal	2021-06-22 17:34	2021-06-22 17:34-18:02

```
Source File: 0000000000000358.tracev3

Date: 2021-06-18 05:37:28 (ADT)

Timestamp: 1624005448637132042

Message Type: Default

Message: Deleting DCIM/100APPLE/IMG_0070.MOV [0xb773dda1a8b83e4a <x-coredata://30230C52-57

Process Name: assetsd

Process Path: /System/Library/Frameworks/AssetsLibrary.framework/Support/assetsd

Sender Name: PhotoLibraryServices

Sender Path: /System/Library/PrivateFrameworks/PhotoLibraryServices.framework/PhotoLibraryServices

Offset: 1260964

Subsystem: com.apple.photos.status

Category: PhotosStatus

PID: 123

UID: 501
```

Fig. 6.72   iPhone X unified log entry from /private/var/db/diagnostics/Persist/0000000000000358.
tracev3

private/var/mobile/Media/DCIM/100APPLE/

The removal of the photos on 2021-06-18 was confirmed through analyzing the unified logs. Unified log entries located in /private/var/db/diagnostics/Persist /0000000000000358.tracev3 confirmed the removal. One of the entries is shown in Fig. 6.72.

The full field content of the message in Fig. 6.72 was: Deleting DCIM/ 100APPLE/IMG_0070.MOV [0xb773dda1a8b83e4a F4F479D8-70B9-4B03-9851-F47DE52F754D] (created on Tue May 18 13:00:08 2021).

The same activity was confirmed to have taken place on 2021-06-19, found in the unified log located at /private/var/db/diagnostics/Persist/0000 00000000035c.tracev3. The log entries are shown in Fig. 6.73.

The full field content of the message in Fig. 6.73 was:

Deleting DCIM/100APPLE/IMG_0094.MOV [0xb773dda1abb83e4a < x-coredata://30230C52-572A-4958-B99F-DF733022B03E/Asset/p94 > 8AEEEF9C-E977-4CE1-A74F-6E1B0697FF42] (created on Wed May 19 19:23:11 2021).

/private/var/mobile/Media/PhotoData/

An   XPC   (Cross   Process   Communication)   activity   called com.apple.quicklook.cloudThumbnailDatabaseCleanup was found to be responsible for the removal of the thumbnail images from the iPhone X. The full deletion activity was located in three different log files:

- /private/var/db/diagnostics/Special/0000000000000141.tracev3;
- /private/var/db/diagnostics/Persist/0000000000000360.tracev3 and
- /private/var/db/diagnostics/Persist/0000000000000358.tracev3.

One of these log entries is provided as an example in Fig. 6.74.

The same activity was confirmed to have taken place on 2021-06-19 for the removal of the thumbnail images on that date, however the log entries were now

```
Source File: 000000000000035c.tracev3
Date: 2021-06-19 05:37:32 (ADT)
Timestamp: 1624091852507127875
Message Type: Default
Message: Deleting DCIM/100APPLE/IMG_0094.MOV [0xb773dda1abb83e4a <x-coredata://30230C52-57
Process Name: assetsd
Process Path: /System/Library/Frameworks/AssetsLibrary.framework/Support/assetsd
Sender Name: PhotoLibraryServices
Sender Path: /System/Library/PrivateFrameworks/PhotoLibraryServices.framework/PhotoLibraryServices
Offset: 1260964
Subsystem: com.apple.photos.status
Category: PhotosStatus
PID: 123
UID: 501
```

Fig. 6.73 iPhone X unified log entry from /private/var/db/diagnostics/Persist/000000000000035c. tracev3

```
Source File: 0000000000000141.tracev3
Date: 2021-06-18 05:37:07 (ADT)
Timestamp: 1624005427823762667
Message Type: Default
Message: com.apple.quicklook.cloudThumbnailDatabaseCleanup:EA096F:[
 ] sumScores:52.520000, denominator:52.520000, FinalDecision: Can Proceed FinalScore: 1.000000}
Process Name: dasd
Process Path: /usr/libexec/dasd
Sender Name: dasd
Sender Path: /usr/libexec/dasd
Offset: 270344
Subsystem: com.apple.duetactivityscheduler
Category: scoring
PID: 104
UID: 501
```

Fig. 6.74 iPhone X unified log entry from /private var/db/diagnostics/Special/0000000000000141. tracev3

located in /private/var/db/ diagnostics/Special/0000000000000147.tracev3, shown in Fig. 6.75.

```
Source File: 0000000000000147.tracev3
Date: 2021-06-19 05:37:10 (ADT)
Timestamp: 1624091830388548500
Message Type: Default
Message: com.apple.quicklook.cloudThumbnailDatabaseCleanup:0BDB70:[
 ] sumScores:51.520000, denominator:52.520000, FinalDecision: Can Proceed FinalScore: 0.980960}
Process Name: dasd
Process Path: /usr/libexec/dasd
Sender Name: dasd
Sender Path: /usr/libexec/dasd
Offset: 270344
Subsystem: com.apple.duetactivityscheduler
Category: scoring
PID: 104
UID: 501
```

Fig. 6.75 iPhone X unified log entry /private/var/db/diagnostics/Special/0000000000000147. tracev3

/private/var/mobile/Library/

With the given modified time of 2021-06-20 03:00, no activities in the timeline nor unified logs could pinpoint a cause for the removal of the images removed from this location in the file system.

6.6   Discussion and Analysis

6.6.1   Comparison Analysis

6.6.1.1   Extraction Sizes

The comparison of extraction sizes across four extractions from five different test devices showed both increases and decreases in size. Both of these results were expected, as SQLite vacuuming and the removal of artifacts would cause decreases in data and powering on an iPhone allows the OS to run, in turn causing increases over the course of the extraction phase.

Overall, the iPhone X was the only test device that saw consistent decreases in extraction size, while the iPhone SE saw consistent increases. The iPhone 5 s and iPhone 7 extractions increased 2 h versus 24 h, but then saw consistent decreases 24 h versus 72 h and 72 h versus 7 days. Alternatively, the iPhone 7+ saw the opposite with an initial increase in extraction size from 2 to 24 h and then consistent

Table 6.6 Noted increases and decreases in data extraction sizes from all test devices

iPhone	2 h versus 24 h	24 h versus 72 h	72 h versus 7 days
SE	Increase	Increase	Increase
5S	Decrease	Increase	Increase
7	Decrease	Increase	Increase
X	Decrease	Decrease	Decrease
7+	Increase	Decrease	Decrease

decreases in extraction size from 24 to 72 h and 72 h to 7 days. These observations are demonstrated in Table 6.6.

In comparing each device against the other, the iPhone X saw biggest overall change in extraction size with decrease of 1,124,936 KB, and the iPhone SE saw the largest overall increase of 418,969 B. In putting all of the device data together, there was an overall loss of 1,193,500 KB and overall gain of 483,472 KB.

6.6.1.2 Artifacts of Interest

Conducting the comparison analysis of the artifacts of interest from all iPhone extractions provided vastly different results. Call Logs were the only artifact that saw no changes at all. The Log Entries and Videos only saw decreases, whereas the Application Usage Logs, Chats, Instant Messages and Images experienced both increases and decreases. These changes are demonstrated in Table 6.7.

To illustrate the changes in artifacts of interest by device, Table 6.8 is provided.

The only changes in artifacts that had an expected result were the changes in the Application Usage Logs and Log entries. The researcher was aware, through previous experience and training, that the knowledgeC.db database operates on a first in last out basis and only keeps entries for a certain number of days. By virtue of the OS running on the test iPhones, new entries would be created by running applications

Table 6.7 Noted changes in the artifacts of interest from all test devices by extraction interval

Artifacts of interest	2 h versus 24 h	24 h versus 72 h	72 h versus 7 days
Application usage log	Both	Both	Both
Call log	n/a	n/a	n/a
Chats	Both	Decrease	n/a
Instant messages	Both	Decrease	n/a
Log entries	Decrease	Decrease	Decrease
Images	Both	Both	Decrease
Videos	n/a	Decrease	Decrease

Table 6.8 Noted changes in artifacts of interest by device

Artifacts of interest	Increase	Decrease	Both
Application usage log	iPhone 7+	iPhone X	iPhone 5s
Call log	n/a	n/a	n/a
Chats	n/a	iPhone 5s	iPhone SE
Instant messages	iPhone 5s	iPhone SE, iPhone 5s	n/a
Log entries	n/a	iPhone SE, iPhone X	n/a
Images	iPhone SE, iPhone 5s	iPhone 7, iPhone X	n/a
Videos	n/a	iPhone X	n/a

and stored in this database. In addition, the lack new Log Entries was expected given that the test iPhones were not in use nor connected to any networks. Which would cause the types of entries stores in the log entries source databases (interactionC.db and DataUsage.sqlite) would be populated with new entries.

The findings of the comparison analysis demonstrated that every test device and every extraction experienced changes. No two extractions were the same size, and each extraction saw an increase or decrease in reported artifacts of interest. Throughout the comparison analysis, the one commonality was change.

This phase of the research also demonstrated that in comparing the total number of artifacts from one extraction to the next, that true number of increases or decreases were not captured until the artifacts of interest were identified in the parsed extractions. As an example, if the 2 h extraction reported 55 log entries, and the 24 h extraction reported 52 log entries in, the assumption would be that 3 entries were lost. However, in identifying and tagging the lost log entries in the extractions, it was learned that there were actually 59 log entries removed from the 2 h extraction, and 4 new entries in the 24 h extraction (Table 6.8).

6.6.2   iOS Application and Database Analysis

The analysis conducted for this portion of the research included comparing and analyzing reported artifacts, SQLite databases, tables and entries, iOS file system locations and '.plist' files.

Four of the five test iPhones reported decreases in their artifacts of interest, and in each of those devices, the removal of data was confirmed. In cases where artifacts were associated to a database's write ahead log, the removal of artifacts was verified in the write ahead log as well as the main database file. This additional analysis was conducted to ensure that artifacts in the write ahead log were actually removed, and not simply written into the main database file. One of the challenges noted in conducting this type of analysis, was the inability to confirm increases or decreases in artifacts when there is no additional data or a source file path or database on which to conduct the analysis. The reason for the reporting of these artifacts by the forensic analysis software could be attributed to a number of factors, however this falls far outside the scope of the author's research and is only being mentioned here to acknowledge hundreds of artifacts that were not able to be verified.

A somewhat surprising outcome that was further analyzed in this phase of the research, were the increases in artifacts of interest, especially considering that these test devices were not connected to any networks and were only interacted with by the researcher for the purposes of performing the data extractions. In total, three of the five test iPhones reported increases. Although the data loss is the focus of this research paper, if the reported increases were able to be associated to a native iOS application, they have been included in the research outcomes.

Of the five test devices, only one reported no data loss at all (the iPhone 7+) and only one device saw some of its artifacts excluded from the research as they were associated to an application fell outside of scope (the iPhone 7).

To summarize all of the quantitative findings from the iOS application and database analysis conducted, Table 6.9 has been provided, which presents a per device summary of all of changes in the artifacts of interest including those that were able to be confirmed, not confirmed (meaning the increase or decrease was able to be disproved through analysis) unable to be confirmed (i.e. no source database to verify was available) or excluded (outside the scope).

In total, across all devices the number of confirmed removed artifacts included:

- 2017 application usage log entries;
- 397 image files;
- 43 chats; and
- 267 log entries;
- 1296 instant messages
- Total loss: 4020 artifacts.

The number of new confirmed artifacts across all devices included:

- 5 images
- 22 application usage log entries
- 36 chats
- Total gain: 63 artifacts.

The number of unconfirmed artifacts across all devices included:

- 26,164 instant messages
- 72 chats
- Total unconfirmed artifact changes: 36,607.

The number of artifacts unable to be confirmed due to no source file across all devices:

Table 6.9 Verified, non-verified and excluded artifacts of interest by iPhone

iPhone	Confirmed ✓	Not confirmed ×	Unable to confirm	Excluded
SE	− 42 chats		− 77 instant messages	
	− 102 log entries			
	+ 2 images			
	+ 36 chats			
5S	+ 13 application usage logs	− 72 chats	− 279 chats	
	+ 1 chat	+ 1 instant message	− 10,853 instant messages	
	− 1296 instant messages	− 23,591 instant messages		
	− 909 application usage log entries	+ 572 instant messages		
	+ 3 images			
	− 1 chat			
7	− 44 images			5 images
X	− 1108 application usage log entries		− 1 log entry	
	− 67 videos			
	− 353 images			
	− 165 log entries			
7+	+ 9 application usage log entries			

- 279 chats
- 10,853 instant messages
- 1 log entry
- Total unable to confirm artifact changes: 11,133.

The number of artifacts excluded across all devices:

- 5 images
- Total excluded artifacts: 5.

6.6.3   Timeline and iOS Analysis

Using the folder or file's modified time as an anchor point to try and find a reason or cause for the increase or decrease in artifacts was more effective when it related to a folder path than a database. With the number of daemons, applications and background processes running at any given time in iOS, the modified time could change a number of times after data is removed, making it far more challenging to pinpoint what occurred.

One thing each device had in common, was the presence of the VACUUM command being used as well as PRAGMA statements invoking the auto-vacuuming to take place. This information was not as easy to find as originally thought. While the iOS unified logs are a good source of information, they appear to be somewhat designed to not divulge all information.

In four of the five iPhones, changes in log entries associated to the knowledgeC.db, interactionC.db and DataUsage.sqlite occurred. Although a cause for any of the reported increases and decreases could not be found, all of these databases had modified times that coincided with the time that the iPhone was being extracted (Table 6.10).

6.7   Summary

The research conducted in this paper has proven that data loss occurs when the extraction of a seized iPhone is delayed, which was demonstrated by a total loss of 1.19 GB of extraction data and the loss of 4020 artifacts. Based on the research findings, and in conducting multiple extractions at different timed intervals, a correlation was able to be sown in the amount of data loss and the length of the extraction delay in that as one increased, so did the other.

With 4083 confirmed changes in the number of reported artifacts across all five test devices, coupled with the fluctuations in phone extraction sizes, the research has demonstrated that an iPhone that must remain in an AFU or Hot state is a volatile container for the data it contains.

In revisiting the ideas proposed in the Background Concepts of this chapter, the research results do support the first proposed solution of extracting a device to place the data into a stable container in lieu of conducting a search incidental to arrest. The totality of the research presented can also be used to support an exigent circumstances framework.

Future Research Possibilities

With the sheer number of devices and applications available, the possibilities for future research become almost endless. As more and more devices turn to a file based encrypted platform, the need to keep devices in an AFU or Hot state in order to be able to extract their data will likely continue to expand well into the foreseeable future.

Since the research conducted in this paper only focused on a specific number of applications, an expansion to include other applications that are not native to iOS would provide an even better understanding of the volatility of data on an iPhone that must be left powered on in order to have its data extracted.

Other areas to expand on would be to include newer and more recently used devices, and to use different extraction timelines. It would not be uncommon for a

Table 6.10 Comparison of changes in log entries

iPhone	Cause of changes found ✓	Cause of changes not found ✗
SE	com.apple.mobileshow	sms.db interactionC.db
5S	sms.db	Photos.sqlite knowledgeC.db
7	/private/var/mobile/Library/Caches/com.apple.MobileSMS/Previews/Attachments/	
X	/private/var/mobile/Media • /DCIM/100APPLE/ • /PhotoData/Thumbnails/*.ithmb • /PhotoData/Thumbnails/V2/DCIM/100APPLE/ • PhotoData/Metadata/DCIM/100APPLE /private/var/mobile/Media/Thumbnails/V2/DCIM/100APPLE /private/var/mobile/Library • /Caches/com.apple.MobileSMS/Previews/Attachments/ • /com.apple.MobileSMS/BrowserSnapshots/ • /SMS/Attachments/	knowledgeC.db interactionC.db DataUsage.sqlite
7+		knowledgeC.db

device to have to wait a few weeks or even longer before it was extracted due to a number of factors, including lack of device support, laboratory backlogs, etc. The ultimate goal in any future research would be to provide research data to better inform law enforcement on what occurs when a seized device is not extracted as soon as possible.

Other possible research that could build on the research presented in this paper would be to delve into the world of Android devices and conduct the same type of research. Alternatively, conducting more focused research into SQLite databases and the vacuuming that takes place in the background specifically as it pertains to iOS.

References

1. Canadian Radio-Television and Telecommunications Commission. (2018). *Consumer Monitoring Report* [Online]. Available: https://crtc.gc.ca/eng/publications/reports/policymonitoring/2018/cmr1.htm. Accessed July 19, 2021.
2. Statistics Canada. (2018). *Canada at a Glance 2018* [Online]. Available: https://www150.statcan.gc.ca/n1/pub/12-581-x/2018000/pop-eng.htm. Accessed July 19, 2021.
3. Aouad, L., Kechadi, M.-T., Trentesaux, J., & Le-Khac, N.-A. (2012). An open framework for Smartphone evidence acquisition. In P. Gilbert & S. Sujeet (Eds.), *Advances in digital forensics VIII* (pp. 159–166). Springer, 8p. https://doi.org/10.1007/978-3-642-33962-2
4. Faheem, M., Kechadi, M.-T., & Le-Khac, N.-A. (2015). The state of the art forensic techniques in mobile cloud environment: A survey, challenges and current trend. *International Journal of Digital Crime and Forensics, 7*(2), 1–19. https://doi.org/10.4018/ijdcf.2015040101
5. Woodburn, T. F. X. (2017). *Digital forensics and domestic terrorism: An investigative study into the Boston Marathon bombings and San Bernadino shootings* [Online]. Available: https://www-proquest-com.ucd.idm.oclc.org/docview/1957428386?https://www.proquest.com/socialsciencepremium?accountid=14507&pq-origsite=summon. Accessed December 21, 2021.
6. Apple Inc. (2021). *Apple platform security guide—Passcodes and passwords*, 18 Feb 2021 [Online]. Available: https://support.apple.com/en-ca/guide/security/sec20230a10d/web. Accessed July 05, 2021.
7. Jordan, R. V. (2016). *SCC 27 [2016] 1 S.C.R. 631* [Online]. Available: https://scc-csc.lexum.com/scc-csc/scc-csc/en/item/16057/index.do. Accessed February 18, 2022.
8. SQLite. (2021). *About SQLite* [Online]. Available: https://www.sqlite.org/about.html. Accessed July 31, 2022.
9. SQLite. (2021). *Database file format* [Online]. Available: https://www.sqlite.org/fileformat.html. Accessed December 22, 2021.
10. SQLite. (2021). *Temporary files used by SQLite* [Online]. Available: https://www.sqlite.org/tempfiles.html. Accessed December 22, 2021.
11. SQLite. (2021). *VACUUM* [Online]. Available: https://www.sqlite.org/lang_vacuum.html. Accessed August 04, 2021.
12. SQLite. (2020). *SQLite—PRAGMA statements* [Online]. Available: https://www.sqlite.org/pragma.html. Accessed July 22, 2021.
13. Apple Inc. (2022). *App store—Apple* [Online]. Available: https://www.apple.com/ca/app-store/. Accessed January 25, 2022.

Chapter 7
IoT Database Forensics—A Case Study with Video Door Bell Analysis

Jayme Winkelman, Kim-Kwang Raymond Choo, and Nhien-An Le-Khac

7.1 Introduction

The use of smart devices has grown enormously in recent years. Many people want to use a smart device in and around their houses. When purchasing a new device, or when replacing an old device. More and more devices are connected to each other, with the possibility of data being exchanged between the devices. This is possible without any user activity. We refer to this as the Internet of Things (IoT). One of the smart devices that occur more is the smart doorbell also named a video doorbell. A video doorbell is an internet-connected doorbell that notifies the smartphone or other electronic device of the home owner when a visitor arrives at the door. It activates when the visitor presses the button of the doorbell, or alternatively, when the doorbell senses a visitor with its built-in motion sensors. The smart doorbell lets the home owner use a smartphone app to watch and talk with the visitor by using the doorbell's built-in high-definition infrared camera and microphone [1]. Lots of people buy a video doorbell to increase security and or for prevention. The result of more video doorbells is that the owners are consciously or unconsciously making (video) recordings of people or situations in front of their houses or company buildings. In the context of digital forensics, this can lead to important evidence or a clue to solve a case. Not surprisingly, the investigators of the investigation teams are trying in various ways to use this information. With more and more video doorbell devices are showing up on crime scenes, it brings up more challenges to the daily work of a forensic investigator. This development forces the Police forces to adjust their mindset and also to change their forensic approach.

Although multiple researches have been carried out on network and IoT forensics, little attention has been employed on how digital forensic techniques can be used to conduct digital forensic investigations into the video doorbell. There is very little research in literature on the video doorbell forensics because of many challenges: (i) identifying necessary pieces of evidence from the video ring doorbell; (ii) collecting and analyzing the evidence from the different environments in the video doorbell eco system; (iii) security of the IoT devices like data encryption and cloud storage; (iv) taking into account legal framework when using video doorbell artifacts.

Therefore, in this chapter, we present a forensic approach for examining the video door bell. We illustrate how to collect and analyze the data from a video doorbell; how to collect network traffic and extract data from the smartphone application of the video doorbell. We demonstrate the examine opportunities of the doorbell and obtain data from the cloud. We are also looking at the legal options in the investigation. We evaluate our approach with the examination of two most popular video door bells: the Google Nest Hello video doorbell and the Ring pro video doorbell, and compare to some of the results of former forensic researches. The main contribution of this chapter can be listed as follows: (i) Identifying an approach to extract the video doorbell artifacts including device forensics, network forensics and mobile forensics; (ii) Forensic acquisition and analysis of two popular video doorbells: Ring Pro and Google Nest; and (iii) Identifying Legal practices to get information from the video doorbell.

The rest of this chapter is organized as follows: Sects. 7.2 describes the research background and the review of related work in literature. Section 7.3 describes the problem statement. We describe our proposed approach for video doorbell forensics in Sect. 7.4. We then evaluate our approaches using different test cases in Sect. 7.5. Finally, we present our conclusion and future work.

7.2 Related Work

In this section we review related research in literature for the IoT forensics. An IoT system integrates various sensors, objects and devices that are connected to the internet and other networks. These devices are capable to connect, communicate and exchange data with each other, without any user activity. Several years ago investigators examined lots of computers. These days they see more smartphones, IoT devices like smart wearable devices, smart home devices etc. The way IoT devices store their information is also different as more and more data is stored online in the cloud. The IoT is presently a hot topic with the constant grow of smart devices in our daily world. The growth of smart devices draws attention from both academic institutions and businesses. Recently, many studies of IoT forensics have been conducted in literature [2–5]. These researches varied from the suggestion of IoT forensic models [3–8] to the forensic analysis of specific devices such as Amazon Echo [4], smartwatch [9], smart sensors [6], etc. Challenges of IoT forensics are also mentioned in [9, 10]. Authors in [10–13] also presented comprehensive surveys on IoT forensics.

Various IoT devices were investigated, as well as the network techniques and protocols. However, due to the enormous growth of IoT devices, there is still much to explore in this field regarding the forensic approach. Search and seizure is an important step in any forensics examination. Nonetheless, detecting presence of IoT systems is quite a challenge considering these devices are designed to work passively and autonomously. There are countless devices that have not yet been researched for suitable approaches to identify, acquire and analyze artefacts [2, 3, 5]. Many studies of IoT devices have shown that there is no standard approach to investigate IoT devices [2–5, 10–13]. Yet, to the best of our knowledge, there is no fully research on the video doorbell forensics that takes into account the extraction and analysis of video doorbell's data in it eco-system.

7.3   Why Video Doorbell Analysis?

The video doorbell is a relative new device in the world of digital forensics. Not much is known about the smart doorbell. The possibilities that provide us evidence. And situations where we have to pay attention because of a video doorbell. Video doorbell devices are connected to the internet with a wireless connection. They are able to communicate with other devices for example with the mobile application on the smartphone and/or with a home assistant and/or cloud service and the Chime. When there is activity in front of the camera the camera starts recording gives a signal there is motion and the facial recognition could recognize a person. When you are not at home the device could tell the person a pre-recorded message without human interaction. In video doorbell eco-system, both network traffic and the stored data are encrypted. The data will be stored in the cloud. A video doorbell is not working without an internet connection. And some of the options of the devices only work in combination with a subscription. Besides the video doorbell also the connected devices like a smartphone with application could provide important evidence these are secured and encrypted and bring other challenges. The video doorbell proper functioning depends on a good network when there is no connection data loss or no notification could be the result. In case of an investigation the data probably would not be found on the video doorbell. The connected devices possibly bring the investigator more opportunities. The connected devices like the cloud storage, the smartphone or other connected smart devices within the network.

Hence, the research objective for this chapter is to examine the possibilities to investigate the video doorbell and its eco-system to extract evidence from the video doorbell. We aim to answer the following research questions: What are the forensic possibilities of a video doorbell? Can we (forensically) acquire images or videos made by the video doorbell? What traces of use can be extracted from the device or the connected network? What information to identify a video doorbell, can be extracted from a connected Wi-Fi network? What are the possibilities of the smartphone application of the video doorbell? What are the legal methods to get information from the video doorbell?

7.4   Methodology

To retrieve information about the video doorbell, in this chapter we are going to examine the following 4 areas within the IoT forensics: (i) Examine the device and perform experiments with the video doorbell; (ii) With network forensics we are going to try if we could discover a video doorbell within a network and examine if the data between the video doorbell and the cloud or mobile phone is encrypted; (iii) Extract information from a mobile phone with mobile forensics. To investigate what kind of information we could extract from the mobile application; (iv) Finally, we investigate the possibilities within the cloud and the connected account. How and what information could we extract from the cloud. We investigate what kind of information is needed to enter the cloud and if we could extract information from the cloud automatically.

7.5   Experiments and Evaluation

7.5.1   Experimental Platforms

The test environment consists of the equipment listed in Table 7.1. The video doorbells and the used mobile phones within the testing environment where connected with a wireless router. The router was connected to the internet and was able to run *tcpdump* to collect the network data. The used laptop was connected by wire. For each experiment the same (wireless) router with *tcpdump* and internet connection was used.

Ring video doorbell and iPhone SE/SamSung S5 both connected wireless to the router on the test network. When the Ring application is started on the mobile phone a DNS request on port 53 is performed. The Ring doorbell also make use of UDP port 53 to connect to DNS. Live view uses TCP destination port 15,064 and on the client UDP ports 30,000 and 30,002. All other connections between the doorbell and iPhone SE/Samsung S5 goes directly to the cloud by port 443 TLS (Fig. 7.1).

Table 7.1 Equipment and version of experimental devices

Hardware	Software	Item
Ring Pro	1.2.9	1
Google Nest Hello	4,110,019	1
Samsung S5	6.01—Android	2
iPhone SE	13.3.1—iOS	2
Laptop	OS Microsoft Windows 10	4
Pineapple Tetra	Open WRT 19.07.2	3

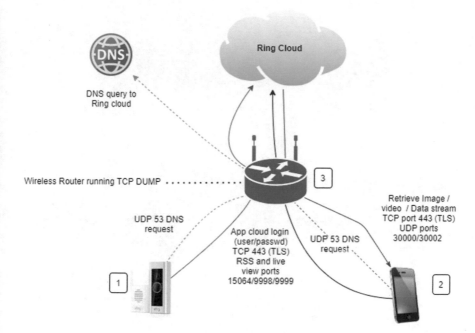

Fig. 7.1 Ring video doorbell with a connected iPhone SE (or Samsung S5)

Ring video doorbell and the Windows laptop connected wireless to the router on the test network. When the Ring website is opened up in a browser, a DNS request on port 53 is performed. The Ring doorbell also uses UDP port 53 to connect to DNS. The connections port 80 to start the website. All communication between the laptop and cloud make use of TCP port 443 TLS (Fig. 7.2). The experiments with Google Nest video doorbell are using the same configuration as Ring video doorbell.

7.5.2 Investigation of the Video Doorbell

The Ring pro doorbell needs a working internet connection to connect to the cloud and to save the recordings. The led status tells if the device is active and if someone is connected. On the Ring pro device, we can find the MAC address of the device. This MAC address must be used when data or information, like an email address and maybe a phone number and payment information are stored will be claimed from Ring Company. The recorded videos are stored 30 days in the cloud. To stop the device cut of the power of the device. No physical options on Ring pro to access the recordings.

The Google Nest Hello doorbell needs a working internet connection to connect to the cloud and to save the recordings. The led status indicates the status of the Nest Hello active and or someone is connected. The Nest Hello doorbell records 24/7

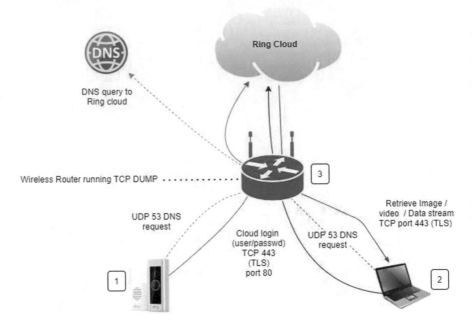

Fig. 7.2 Ring video doorbell with a connected laptop

audio and video. The motion and actions will be saved in the cloud and could be claimed from Google LLC. Also the Google account and all connected information to the device and the account could be claimed from Google LLC. The MAC address and information of the device are only available with use of the Nest application or extract from the connected network. To stop the device working cut of the power of the Nest Hello. No physical options on Nest Hello to access the recordings. Only the chip off method could exclude if there is stored data locally within the video doorbell.

When a chip-off will be performed on the video doorbell the memory chip will be extracted from the device within a lab environment. The data will be imaged and could be investigated with forensic tools to examine the data if there is data stored. This data could also be encrypted if so this option would not provide us any information.

7.5.3   Network Examination

7.5.3.1   Network Scan

Within the test environment we performed some tests to look at the options to trace the video doorbells. When performing a network scan with a network sniffer. We start a scan not connected to a wireless network. The scan for available Wi-Fi networks and wireless clients is performed to collect mac addresses from devices. The Ring doorbell could be recognized when the device is in setup mode because of the showed SSID (Ring and the last characters of the mac address.) When a functional Ring already is configured and connected to a network, the device will not send the SSID. When you look at the collected mac addresses you cannot filter out the doorbell on the mac address. When you look up the vendors of the collected Ring mac address, Universal Global Scientific Industrial Co. Ltd is showed. The Ring Chime will show the vendor Texas instruments. When you look up the mac address of the Google Nest Hello the vendor Nest Labs Inc will be showed.

When the investigator connects to the same wireless network as the Ring pro and Google Nest Hello and perform a network scan to discover connected devices on the network. The Ring pro doorbell will be found on the network. The Ring pro is found with the name combined with the last digits or letters of the mac address from the device. For example, this could be RingPro-00.lan when the mac address ends with 00. The Nest Hello doorbell will be found on the network with the Name Nest-Hello and last 4 digits of its mac address for example Nest-Hello-00e0.lan when the mac address ends with 00:e0.

It is not possible to connect to the Ring pro doorbell directly. The only connection to the doorbell is through the internet. The device sets up a connection to the cloud. And the other device that wants to connect to the doorbell, for example a smartphone setup his own connection to the cloud. Now both devices could communicate with each other.

During the experiments with the video doorbells, the data from the connected network, was captured with *TCPdump*. The captured data will be saved to a *pcap* file. These *pcap* files are going to be analyzed with the program Wireshark. For each experiment we create a new *pcap* file so we could compare the data and the results of the *pcap* files afterwards.

7.5.3.2 Ring Pro Video Doorbell Activity

First we start with the tests of activity on the Ring pro video doorbell. The doorbell is connected to the network and an account is configured. During the tests the Ring plan is active by 30-day trial so all the data will be stored. All the recorded videos will be deleted at the end of the testing period. During the tests with ringing the doorbell it is performed with motion on and off to create data and to be able to analyze if there is difference within the captured data. All the tests were performed multiple times to see if there is a pattern in the data.

The Ring pro doorbell connects to the internet even when there is no activity in front of the doorbell it connects to the cloud. The doorbell connects to several cloud servers of Amazon and Cloud flare. Besides these servers there are multiple connections to Ring related domains. Ring pro video doorbell uses the local network to connect to the internet, it does not expose any services to the local network. The connections are setup with HTTPS and TLS all encrypted network traffic.

These protocols are also used within the data TCP, DNS and UDP. After analyzing al the network data form al the experiments, the following list of subdomains of resolved addresses from Ring.com are found: *alerts.ring.com.cdn.cloudflare.net; api.ring.com.cdn.cloudflare.net; billing.ring.com; controlcenter.nw.ring.com; es.ring.com; fw.ring.com; fw-snaps.ring.com; nh-mobile-config.ring.com; ps.ring.com; oauth.ring.com; Link.verify.ring.com; site-nac.ring.com; az.ring.com; share.ring.com; static.ring.com; account.ring.com; alerts.ring.com; ring-untranscoded-videos.s3.amazonaws.com.*

Some of the names of the subdomains could be a clarification what the target is and why it is used for. Billing.ring.com could be a sort of check if a plan is activated on the account. Several servers are used as API from Ring. Account.Ring.com and control center are options within the Ring account. There is no readable data captured from the Ring device as all data is encrypted.

7.5.3.3   Google Nest Hello Video Doorbell Activity

The doorbell is connected to the network and an account is configured. All the recorded videos will be deleted at the end of the testing period. All the tests were performed multiple times to see if there is a pattern in the data.

Within the captured data the following protocols are showed up. TCP, UDP and TLS. All the data from and to the Google Nest Hello doorbell is encrypted. There is no readable data. There are multiple DNS requests to the following sub domains from nest.com. Resolved addresses Nest are listed as follows: *apigw.production.nest.com; weave-all-regions.production.nest.com; time.nest.com; logsink.home.nest.com; nestauthproxyservice-pa.googleapis.com; webapi.camera.home.nest.com; NestLabs_6; home.nest.com;* www.google apis.com; *apps-weather.nest.com; googlehomefoyer-pa.googleapis.com; firebaseremoteconfig.googleapi-s.com; nexusapi-eu1.camera.home.nest.com; czfe150-front01-iad01.transport.home.nest.com; nest.com; store.nest.com.* The Google Nest Hello connects to the cloud and to several online servers of Google Nest. There is no local network traffic that was traceable or readable. The Nest Hello connects to the internet and to time.nest.com to sync time. It also gets the weather information online from apps-weather.nest.com. It is not clear how the data is synced and stored online and uses multiple online servers. The transport of data is fully encrypted.

7.5.3.4   Discussion

Both Ring pro and Google Nest Hello doorbell show the same result. As expected, research has shown in the data all the network traffic is fully encrypted. No information could be traced. No video streams can be watched from capturing network data. We can see network requests to online cloud servers. We see several DNS requests to servers from Ring or Nest. This hints that there is a connected device but you cannot trace what device is communicating and how. We can also conclude both devices don't communicate directly to devices on the local network. All the communication goes by the cloud environment. Because both devices work with certificates there is also no man in the middle attack possible. When there is activity on the Ring pro or Google Nest Hello device there is also network traffic to the internet but there is no pattern and all is encrypted. When sniffing for wireless networks and clients we could not find indications about Ring. When you look up the vendors of the collected Ring mac address, Universal Global Scientific Industrial Colt is showed. The Ring Chime will show the vendor Texas instruments. We could find an indication about the Nest Hello When you look up the mac address of the Google Nest Hello the vendor Nest Labs Inc. will be showed. When there is access to the same network and performing a network scan the Ring pro and Google Nest Hello will be recognized within the network and the scan results. The encrypted connection to the cloud goes on port 443 with the TLS protocol. Within the experiments we captured the data on the router. The only way we could see what kind of data is send we should look at the device or at the endpoint in the cloud. Within the test environment we tried if the device is reset or factory default if we have access to the data but this won't provide us access to the stored cloud data. When we shut down the connection to the internet the device and the connection to the cloud are not available. Could we have access to the network data when we could bypass the security of the TLS or when there is a bug in the software could we have access? When we have access to the email client of the connected account that is used to login to the account of the video doorbell that store the data within the cloud we could reset the password and have access to the user environment.

7.5.4 Smartphone Application

7.5.4.1 Ring Mobile Application

In this test, the Ring application was installed on an iPhone SE and Samsung S5, and the account environment online opened within the browser Google Chrome on Windows 10. When an active Ring application is opened, the dashboard is shown. Within this dashboard on top the location is shown, this location is added by the owner, and is not a location added with location service of the device. Event History and the added devices are also shown. When opening the Event History, there is a list of recent activities with the activated time.

When the owner deletes a video it disappears from the event history. There is no information about deleted events or videos. It is only a time line of the current stored activity videos. There is also no log within the application about activity within the application or on the device.

There is an option to share or download the stored videos within the Ring application. This could be done by download the video to the device a smartphone or computer. Or the option to share the video by sending a download link of it. These options are available to share the video from the mobile applications such as Facebook, WhatsApp, Email, Next-door, etc. All the users within the account have these options.

In relation to investigations the Devices and History options are interesting. We can see what connected devices there are active and what activity is stored. The settings of the connected devices and the connected Ring account and the information connected to the account like email and phone number.

The Ring pro device settings could be interesting to know how the connected device was configured. Ring alerts and motion alerts could be on/off. The Device health tab gives information about the Ring pro device and if the Ring device is working properly. The Wi-Fi connection and firmware are checked status is shown, and the MAC address of the device is shown at this location.

The Ring device could sent multiple Ring notifications to the connected users and applications. When motion is detected, when someone pushes the button of the doorbell. Ring Modes there are 3 modes to choose Home, Away or Disarmed. The account user of the Ring pro device could choose if the devices record and sent notifications or disarm the device. When adjusted these setting and change the mode the connected Ring account user receives a notification of the chosen mode of the devices. The Ring notification consists only of text.

7.5.4.2   Google Nest Mobile Application

In this test, the Google Nest application was installed on an iPhone SE and Samsung S5, with the account online within the browser Google Chrome on OS Windows 10.

When open up the Google Nest mobile application the first time you have to add an account a Gmail or nest account. Afterwards the app opens and shows the view of the connected Nest Hello doorbell and the temperature of the added location of the Nest Hello.

When activate the camera view, two options appear to see a live stream or the recorded activity could be watched. The camera of the Google Nest Hello doorbell streams day and night and records the activity within the view of the camera or when a loud noise is received. When a plan is activated there are extra options within the doorbell like facial recognition. During the examination the plan option was not activated. The Nest Aware functions could not be tested and these options are not performed during experiments and examination of the Google Nest Hello doorbell. The doorbell sent notifications to the connected application. Within this notification a snapshot is added. The Google Nest Hello has 2 options when an account user of the Google Nest Hello device is at home or is away the settings could be changed. The home motion could disable the camera or when there is a mobile device is connected and the location is shared the camera could switch automatically between home or away. When the connected phone is near the device the camera switch to home mode and when leaves switches to away mode and records al the motion and activity.

When there is motion in front of the camera the Google Nest Hello is able to filter between people or animals and even has the option to recognize people by facial recognition. When these options are active the device could also activate an audio message to the person in front of the camera without being at home. Within the tests not every motion is triggered. Google Nest Hello gets active by motion, sound or when someone rings the doorbell by pressing the button. Depending on a good network connection to send triggers and notifications. The recorded videos from the Nest Hello are stored to the cloud. When for example other Google devices are at a home like the Google Home these devices could communicate and for example the Google Home could notify and give an announcement when a visitor is in front of the camera and if recognized by the face recognition tells who is in front of the Google Nest Hello.

The quality of the captured stored video could be adjusted to lower the bandwidth the Nest Hello uses. The account settings are in the Nest application and an email address and optional a connected device smartphone is shown. There is also an option to share the device with family and add other users to receive notifications and give access to the stored Nest videos.

The changes of the state of the Nest Hello device, home and away mode are stored in the activity history with day and time. And when a Nest Aware subscription is active the owner could receive a Home report every month. This report contains energy use of the device and safety events. This report will be received at the connected email address of the Nest account.

7.5.4.3 IPhone SE Mobile Forensics

The iPhone SE is connected with the Ufed Touch from Cellebrite [14]. With the Ufed Touch the user data and file system could be extracted from the mobile phone. The data is extracted and stored as an Ufdx file. This file is opened with Ufed Analytics software to analyze the data and generate a report with data an Ufdr this report could be opened with the Ufed Reader.

We look at the data in a timeline format what information could we extract from the used forensic software and the generated timeline are the notifications from the application written in this timeline or when the mobile applications of Nest and Ring are used?

After performing the keyword search performed within the extracted data from the iPhone SE with the keyword terms: Ring and Nest. Several hits on the results of the keyword search on the terms Ring and Nest.

If we look at the results of Ring and Nest locations within the file system with an investigators perspective, the files and folders interesting for the investigation are shown in Fig. 7.3.

Within the log files in the shown folder logs, we found related information about the used Ring pro doorbell. Databases and *plists* files are found in the *com.ring* folder (Fig. 7.4). In the database *linphone_chats.db* we discovered some information about *sip* connections to IP address related to Ring.com and the SIP number and event camera connected. This are some connections to the Ring pro video doorbell from the mobile application. We discovered the account used for Ring, the used

Fig. 7.3 Folder
group.com.ring containing
folders and files

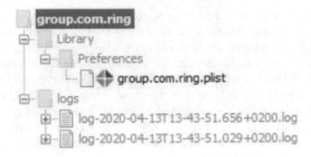

email is found at the *localstate.db* database. Related to the downloaded videos from Ring we found in the database some of the shared links from the Ring videos. The examination of the notifications did not result in any hits of Ring notifications. In the Ufed analytics timeline there were no results of the Ring application.

When we investigate the artifacts of the Nest application in the extracted data. We extract 2 sqlite databases containing important information: *Nest.sqlite* and *Dropcam.sqlite*. The table ZCDBASEDEVICE contains the mac address, model of the device, and serial number of the connected Nest device. The table *ZCDUSERS-ESSION* contains the Nest account and connected email address. *ZCDSTRUCTURE* contains the added postal code of the location of the device. The examination of the notifications did not result in any hits of Nest notifications. There are no results in the Ufed analytics timeline there were no hits on the Nest application.

7.5.4.4   Samsung S5 Mobile Forensics

The Samsung S5 is also connected with the Ufed Touch from Cellebrite [7] to extract data. First we perform a keyword search through the extracted data from the Samsung S5. The keywords Nest and Ring only appear in the list of applications. Both Ring and Nest application are linked to the *localappstate.db*. This database contains the email address of the account that is used at Nest and Ring. There is no readable data that is linked to the mobile applications of Ring and Nest. There are no messages of notifications found on the data. On the SD card from the Samsung were found 2 video files, these files are downloaded from the Ring account. The filename of the video files contains Ring Video and the date and time of the recording.

7.5.4.5   Discussion of Results

To get the best result of extracting information from the mobile application, the best method is to open the physical device with a working internet connection. With a working internet connection there is full access to the cloud and all the stored information. The cloud environment is available and all of the recordings could be viewed within a timeline and downloaded. When there is no connection available

Fig. 7.4 *com.ring* folder in file system with databases and plists

the account information like account name and connected information like email and optional a phone number could be found within the application. Only the cloud environment is unreachable. The MAC address of the device could not be extracted from the mobile application when there is no working internet connection. The MAC address and account information could also be extracted from the data of an iPhone. The database *localstate.db* contain the account information of the Ring account. From the Nest application from the iPhone the information could be extracted from a *Nest.sqlite* database. The biggest challenge is to find the readable data if there will be access to the phone data. When the account information is found this information could be used to receive the recorded videos. This could be claimed by Ring BV or Google LLC whit a subpoena. The account could also be used to access the cloud when the legal opportunities are there.

7.5.5   Securing Data

After investigating the mobile application, we have discovered the options of extracting data from the account from the cloud environment. By using the mobile application, and the option to log in with a computer browser, and share or download the recorded videos.

If it is possible to extract data from the cloud environment without using the mobile application or a computer browser, we could extract the data by using a script. We examine this possibility to see if we could connect to the cloud and extract information and recorded videos. Because there is no subscription to the nest environment we investigate the option to generate information from the Ring API cloud environment.

The different methods have an effect on the result. For example, when sending a video with email or WhatsApp will affect the quality because of compression. Google Nest Hello has several options to save the videos from the different devices shown in below table.

We examine the possibility to locate the connected Ring video doorbell and extract recorded video from the cloud environment with the Ring API using the available python library from https://python-ring-doorbell.readthedocs.io. Now we could connect to the Ring "cloud" account. To login we need the Ring username and password, and we need access to the connected email account to receive a verification code. These credentials create an access token. With the token we could connect to the API. Within the library there are several options to perform to extract information about the Ring pro video doorbell such as the last actions (Fig. 7.5). We can retrieve the location and the connected Wi-Fi network of the Ring pro doorbell. The location of the device, is the location that is filled by the setup not a location retrieved from location service. We are able to create a snapshot from the view of the Ring pro doorbell. It is also possible to extract the latest recorded video or multiple videos from the cloud environment. The downloaded videos contain and display the date and time of the recorded video when play the video. The videos are stored with HD quality.

Ring and Nest both have an API to communicate with. This API brings us an extra method to extract information and secure the videos from the cloud. When the account of the Ring is available and the connected email to retrieve a security code.

Fig. 7.5 Overview last actions of the doorbell extracted with the script

There are several scripts to retrieve information. One of the options is to retrieve videos automatically from the cloud environment of the used account. To use the script and to extract videos from the account we need a username and password. Also could be the possibility that we need a security code that will be generated for extra security. When we have access to the email account that is connected to the used cloud account we could reset the account or change the password. On this way we could have access to.

Another option would be to hack the account to get access this would not be an option in the Netherlands because we don't have a legal option to perform this action. When we don't have the account that is connected to the device we could request the account information with a subpoena when we have the location and mac address of the used device. We need also a connected device like a smartphone to get access to the account.

7.5.6 Law and Guidelines

In some countries we have to deal with the law and regulations when we have an investigation the prosecutor is the leader of an investigation.

Securing data from a network, investigating a network and the extracted data
If we want to perform an investigation on a network. For example, the home network of the suspect or the network of a company we need permission. From the place where a search takes place, it is allowed to perform a network search in an automated online network. Research is done into data stored in that network. If data is found that can reveal the truth, then could be recorded or extracted. We are not allowed to extract data from the cloud when the investigator knows or could know the data is stored abroad outside the borders.

Seizure of a mobile phone, extract the data and analyze the data
We could seizure a mobile phone of a suspect. When we extract the data and want to analyze the data we need permission from a public prosecutor or examining magistrate applied because of the privacy of the suspect. The general seizure authorizations can be found in each country law. When we want to seizure a phone within a house or building of the suspect we need a search warrant.

Extracting data from the cloud environment
We are allowed to extract data from the cloud environment of a suspect when the suspect gives freely permission to access the account. We could ask permission to the examining magistrate. And also need the network search permission.

Retrieve information from Google and Ring
When the prosecutor grants permission the law enforcement seeking to obtain data from Ring must sent the request through the legal and diplomatic channels in its

jurisdiction. Ring will release user information and recorded videos to law enforcement in response to a valid and binding legal request. This could be a valid subpoena,
search warrant, or other court order request. Requests and questions should be sent
to subpoenas@ring.com. Within the request the Account or Mac address must be
added. Nest (Google LLC) will release user information and recorded videos to
law enforcement in response to a valid and binding legal request. The legal request
is sent to Google Ireland Limited by the single point of contact from the police.
The Dutch law, for example offer several options, to retrieve the data and possible
evidence within the legal borders. When the recorded videos of Ring or Nest must be
downloaded, and the owner gives no permission, the only method is to send a legal
request to Ring or Nest. Both companies have legal guidelines and if the legal request
complies with regulations they will cooperate and provide the requested information
or recorded videos from the account of interest.

7.6 Conclusion

This chapter presented the process of securing data recorded by the video doorbells
from Ring and the Google Nest Hello. The connected environment is very important
to investigate when we come across a video doorbell at a crime scene. With help
of mobile and network forensics we could provide important information. We could
collect the information about the used device or the account information. When we
have access to the connected device like the mobile application this would give us
the opportunity to view the recordings stored in the cloud.

Because the video doorbell stores their data within the cloud storage this will bring
us legal issues. We could not create access to the cloud or download the videos from
another country. Without the permission of the owner we need to send a subpoena
to Amazon Ring or to Google and request the recordings in a time frame.

This chapter showed that the video doorbell connects to the cloud. However the
device could cache data locally and when a chip-off is performed, future work should
examine if data is stored on the device itself. Because of the limitation the plan of
Google Nest Hello was not active, the API of the Google Nest Hello was not tested.
The options this would bring could be tested. In addition, the communication between
the video doorbells and other devices like the smart home device could be tested and
analyzed.

References

1. Shu, L. (2013). *Smartbell is the high-tech doorbell that lets you video chat with your visitors.*
   Retrieved July, 2020, from https://www.digitaltrends.com/home/smartbell-doorbell
2. MacDermott, A., Baker, T., Buck, P., Iqbal, F., & Shi, Q. (2019). The Internet of Things:
   Challenges and considerations for cybercrime investigations and digital forensics. *International*

Journal of Digital Crime and Forensics (IJDCF), 12(1). ISSN 1941–6210 Retrieved from http://researchonline.ljmu.ac.uk/id/eprint/9496

3. Sathwara, S., Dutta, N., & Pricop, E. (2019). IoT forensic—A digital investigation framework for IoT systems. In *10th International Conference on Electronics, Computers and Artificial Intelligence (ECAI)* (pp. 28–30). Iasi, Romania, Romania.

4. Li, S., Choo, K. K. R., Sun, Q., Buchanan, W. J., & Cao, J. (2019). IoT forensics: Amazon echo as a use case. *IEEE Internet of Things Journal, 6*(4).

5. Meffert, C., Clark, D., Baggili, I., & Breitinger, F. (2017). Forensic state acquisition from Internet of Things (FSAIoT): A general framework and practical approach for IoT forensics through IoT device state acquisition. In *ARES '17: Proceedings of the 12th International Conference on Availability, Reliability and Security* 2017. https://doi.org/10.1145/3098954.3104053

6. Goudbeek, A., Choo. K. -K. R., Le-Khac N. -A. (2018). A forensic investigation framework for smart home environment. In *17th IEEE International Conference on Trust, Security and Privacy in Computing and Communications*, New York, USA. https://doi.org/10.1109/TrustCom/BigDataSE.2018.00201

7. Hilgenberg, A., Duong, T. Q., Le-Khac, N. A., & Choo, K. K. R. (2020). Digital forensic investigation of Internet of Thing devices: A proposed model and case studies. In N. A. Le-Khac, & K. K. Choo (Eds.), *Cyber and digital forensic investigations. Studies in big data* (Vol. 74). Springer. https://doi.org/10.1007/978-3-030-47131-6_3

8. Alabdulsalam S., Duong, T. Q., Le-Khac, N. A., & Choo, K. K. R. (2022). An efficient IoT forensic approach for the evidence acquisition and analysis based on network link. *Logic Journal of the IGPL*. https://doi.org/10.1093/jigpal/jzac012

9. Alabdulsalam, S., Schaefer, K., Kechadi, T., & Le-Khac, N. A. (2018). Internet of Things forensics—Challenges and a case study. In G. Peterson, & S. Shenoi (Eds.), *Advances in digital forensics XIV. DigitalForensics 2018. IFIP advances in information and communication technology* (Vol. 532). Springer. https://doi.org/10.1007/978-3-319-99277-8_3

10. Alenezi A., Atlam J. G., Alassafi M. O., & Wills G. (2019). IoT forensics: A state-of-the-art review, challenges and future directions. In *4th International Conference on Complexity, Future Information Systems and Risk (COMPLEXIS 2019)*, Crete, Greece. https://doi.org/10.5220/0007905401060115

11. Hou, J., Li, Y, Yu, J., & Shi, W. (2020). A survey on digital forensics in Internet of Things. *IEEE Internet of Things Journal, 7*(1).

12. Yaqoob, I., et al. (2019). Internet of Things forensics: Recent advances, taxonomy, requirements, and open challenges. *Future Generation Computer Systems, 92*, 265–275.

13. Mendez, D., Papapanagiotou, I., & Yang, B. (2017). Internet of Things: Survey on security and privacy. *Information Security Journal*. https://doi.org/10.1080/19393555.2018.1458258

14. https://www.cellebrite.com/en/ufed-ultimate/

Chapter 8
Web Browser Forensics—A Case Study with Chrome Browser

Jacques Boucher, Kim-Kwang Raymond Choo⊚, and Nhien-An Le-Khac⊚

8.1 Introduction

8.1.1 Web Browser Forensics

Internet is widely used in the world and its popularity has grown significantly since 90s of the last century. In 1994, only around 0.04% of the world's population (~25 million users) had Internet access. By the end of 2021, over 53% (~5 billion users) of the world's population had access to the Internet, almost 800,000 new users each day [1]. The use of a software program known as a web browser remains a popular way to access content on the Internet. Consequently, a significant number of activities are conducted daily on the Internet using web browsers. The most popular online browsing activity is to search for information using different search engines. As of March 2021, there are approximate 5.5 billion Google searches per day, the most popular search engine.

Today, the World Wide Web is an integral part of modern life, and web browsers remain a popular means of accessing its content. Browsers are used to search the Internet for content, watch videos, listen to music, e-commerce, online banking, social media activities, and so much more. There are dozens of web browsers available today, however four of them dominate the landscape: Google Chrome, Safari, Microsoft Edge (which is now Chromium based), and Firefox. As of 2022, Google Chrome has the biggest market share with 64%. Safari is around 19%. Firefox and Edge have about 7% of the market. The top browsers are available on both desktop

---

This chapter was written outside of the first author's employment. Any opinions expressed by the author in this chapter are the authors' own personal opinions. They do not represent those of first author's employer.

OS' (Microsoft Windows, Apple OS, Linux) and mobile OS' (iOS, Android). The desktop vs mobile version of these cross-platform browsers tends to use the same artifact storage structure, making your browser analysis forensic skills equally cross-platform. The popularity and usefulness of the Internet has created a new form of crime, "cybercrime", and a new category of criminals, "cybercriminals". Cyber-criminals are exploiting the Internet's convenience to engage in criminal activities worldwide. There are approximately 200,000 websites hacked each day [1], 88% of organizations experienced spear phishing attempts in 2019 around the world [2] and 9.9 billion malware attacks in the same year [3]. Cybercrimes are not only crimes targeting computer hardware and software, and Internet of Things (IoT) devices, but also cyber-enabled crimes such as crimes against children, financial crimes, and terrorism. Such cybercrimes can involve the use of browsers to engage in criminal activities.Cybercriminals have been exploiting the popularity of web browsers for their illegal activities [4]. Web browsers could leave recoverable traces of their activities, hence searching the evidence left by web browsing becomes a crucial task for the digital investigator. Analyzing these traces can lead to varied artifacts including websites visited, keyword search terms used, timestamps of activities, etc. There are many sources of evidence in web browsers (a.k.a. browsers). For instance, the browser's history contains a list of web pages that were visited by a user. Browser cookies, small text files stored on the user's device by the website being visited for future tracking, can yield evidence. Bookmarks/Favourites will contain links that a user opted to save of web pages they found interesting to facilitate navigating to them again in the future. Browser cache contains locally stored copies of web pages that a user visited. These browsing artifacts can reveal the browsing habits of a user and may contain evidence of a user's online criminal activity. These artifacts can facilitate a pattern-of-life analysis of a user, determining the objectives, methods, and activities of a cybercriminal. In web browser forensics, simple browser analysis with the aid of automated tools is not enough [5]. Today, it requires more advanced forensic techniques to extract as much evidence as possible from browser artifacts. When more than one browser is used, artifacts from each must be analyzed and merged into a single timeline to get a more accurate picture of a user's online activity. We will focus on Google Chrome forensic analysis. There are other valuable resources [6, 7] available to assist with the analysis of other browsers. Fortunately for the forensic examiner, most browsers you will encounter use SQLite to store artifacts, allowing skills you will learn in this chapter to be applied to other browsers.

8.1.2 Google Chrome

In this chapter we will look at Google Chrome artifacts found in various Google Chrome databases (SQLite and JSON formats). It is not an exhaustive look at all databases in Chrome, but it will cover the more common ones. Most of the topics in this chapter could be a chapter on their own in a dedicated Chrome Forensics book.

Google Chrome is based on Chromium's code. "The Chromium projects include Chromium and Chromium OS, the open-source projects behind the Google Chrome browser and Google Chrome OS, respectively" [8], because Chrome is based on Chromium, you can search through Chromium's source code at https://source.chr omium.org/chromium as part of your research into Google Chrome. Even if you are not a coder, developer comments in Chromium's source code can help you form an opinion about some of Google Chrome's artifacts and reference the source code to support that opinion.

Google Chrome is not the only browser that is based on Chromium. There are many browsers based on Chromium. Two of the more recognized ones are Microsoft Edge and Opera. Browsers based on Chromium each will tweak the end user experience through changes to the user interface and adding or changing some of the front-end features. But they will share the core artifacts. If you know how to analyse the core Google Chrome artifacts, then you know how to analyze the core Opera and MS Edge artifacts as well (and many others).

The other convenience we enjoy is that the artifacts are the same across different operating systems (Windows, Mac, Linux, Android, iOS[1]). This means that if you know how to analyze core Google Chrome artifacts on a Windows computer, you will be able to analyse them on a Mac or Linux computer, or on a mobile device.

Up until Q2 of 2021, Google Chrome was on a 6-week rapid release cycle. Meaning updates were released every 6 weeks. As of Q3 of 2021 with the release of Chrome 94, it switched to a 4-week rapid release cycle. These releases can contain bug fixes, security patches, feature enhancements, or new features. Because of this, it's possible that some content in this chapter will no longer apply in a future release.

The impact of this rapid release cycle is not only a challenge for forensic examiners. It's also a challenge for forensic tool developers. Staying abreast of changes in each release of Google Chrome is challenging, as Google Chrome is just one of many artifacts a forensic tool must support (think of all the OS artifacts, and all the other applications the forensic tool parses and must remain abreast of those changes as well).

---

[1] Chromium uses the Blink rendering engine. Apple only allows WebKit, its own browser rendering engine, on iOS devices. Thus, all browsers running on iOS use the WebKit rendering engine. This does not appear to impact the artifacts created by the front-end user interface, but it's worth being aware of this fact.

8.1.3   Forensic Tool Gone Wrong

The value of knowing your artifacts to recognize when your forensic tool is not properly decoding something was observed by the author while teaching a class on Google Chrome forensics in the early days of Google Chrome. Students were taught how to manually find and analyze Google Chrome artifacts. During one of the exercises, a student used a well-respected and still widely used forensic tool to answer an assignment question. The student noted that the visit date of a URL that had been visited multiple time was identical for all entries according to the forensic tool. The student called the author over for assistance. The author immediately recognized the problem. The forensic tool was using the last visit field in the URLs table rather than the visit time field in the visits table (more on this when we look at history artifacts in this chapter). A bug report was filed with the tool, and it was corrected very quickly. But it was a great teaching moment that underscores the value of having a better understanding of the artifacts so you can recognize when one of your tools is inaccurately parsing the data and validate it either manually, or with another tool.

8.1.4   Environment Variables

In this chapter the author will reference path to files using environment variables. On a Microsoft Windows computer, you can view your case insensitive environment variables in a few ways. The author's preferred method is to open a command window and type the command "set" to view the variables. You can also see the value for any single variable with the "echo" command. For example, to see the folder for Program Files, you can type: "echo %ProgramFiles" (Fig. 8.1).

Environment variables are ways for applications to know where to find something on a Windows system without worrying if a user installed Windows in a non-default folder for example. Some of the ones you may see in this chapter are the following: AppData, LocalAppData, Username. You can display their value at the command line with the echo command (Fig. 8.2). Of course these are the value of the variables on your system, not on a forensic image.

As a tip, you can use an environment variable in Windows Explorer to navigate to a folder. Opening Windows Explorer and entering %localappdata% in the address bar

Fig. 8.1 %ProgramFiles% environment variable

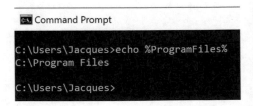

Fig. 8.2 Other environment
variables

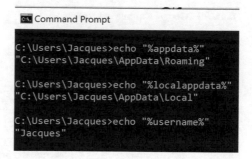

will navigate to the corresponding folder on your system. They can also be used at the command line if you prefer navigating a system that way, e.g., cd %localappdata%.

8.1.5   Epoch Time or WebKit Time

It's important to note that you will encounter both Epoch time stamps and WebKit time stamps (and even some text time stamps) when analysing Google Chrome database files. Some of you may be wondering why WebKit time if Google Chrome is based on the Blink browser engine? The author has not researched the answer, but we do know that Blink is a fork of WebKit [9], the browser engine created by Apple and used in Safari [10]. That may explain the reason for WebKit time stamps. With experience you will recognize the difference between the two. Where applicable, the author will point out which applies when referencing dates in this chapter.

8.2   Chrome Artifacts Folder

Google Chrome's artifacts are at the following path (Table 8.1). The rest of the chapter will provide you with the relative path found at this base path.

Local State

Local State is a JSON [12] file. JSON files are plain text files with key:value pairs. If you are not familiar with JSON files, I encourage you to read up on them. You will encounter them in browser forensics (not just Chromium bases browsers, but also Mozilla Firefox based browsers), as well as other areas of forensic analysis. The author assisted a forensic analyst who had recovered from unallocated a partial sessionstore.js file (Firefox session recovery file) as part of a homicide investigation. Its content was relevant to the investigation, but the examiner was seeking assistance in interpreting it and rendering it in a more legible format. With the author's help, the partial JSON file was fixed by adding a few missing tags so that it could be properly

Table 8.1 Google Chrome data folder

OS	Path to file
Microsoft Windows XP	C:\Documents and Settings\%USERNAME%\Local Settings\Application Data\Google\Chrome\User Data\
Microsoft Windows 10/8/7/Vista	%LocalAppData%\Google\Chrome\User Data/
OS X/Mac OS	~/Library/Application Support/Google/Chrome/
Linux	~/.config/google-chrome/
Android	data\data\com.android.chrome\app_chrome\Default [11]
iOS	%root%\Library\Application Support\Google\Chrome\Default [11]

rendered by a JSON viewing tool of choice. E.g., Notepad++ with the appropriate add-on, Firefox' JSONView add-on [13]. The latter requires that the file has a .json extension to parse it out using the browser add-on.

The file Local State contains many valuable settings, of which only a handful will be covered in this chapter.

8.3 Profiles

Google Chrome allows you to create user different user profiles. This can be helpful if you have a shared family computer with a single sign on for everybody. Each family member can have their own Google Chrome profile so that they have their own bookmarks, browser history, saved passwords, etc. It can also be used by someone wanting to keep things organized by having a profile for work, one for home, and maybe another for their hobby. A digital forensic examiner might use a few different profiles as well. All profiles are stored in subfolders in Google Chrome's artifacts folder covered earlier.

If there is only one profile, the artifacts for that Google Chrome user will be stored in a subfolder called "Default". All subsequent profiles are in subfolders called "Profile 1", "Profile 2", etc., also located directly off Google Chrome's base artifacts folder. If present, each of these profile folders will contain their own user artifacts that you will want to analyse.

This creates a caveat when analysing a system with your favourite forensic tool.

- Does your forensic tool parse all the Google Chrome user profile folders?
- If yes, does it report everything together, or does it identify from which profile the artifacts came from?

During your forensic analysis, you will see the "Default" folder, and the various "Profile #" folders if present by navigating the file system. In the example (Fig. 8.3), we see a total of three profile folders (Default, Profile 1, Profile 2).

Fig. 8.3 User profile folders

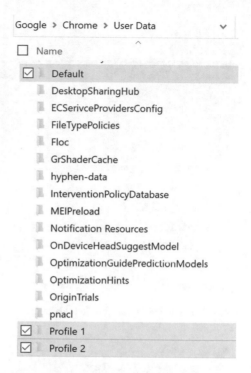

You can also verify how many profiles are on a system by examining the "Local State" file at the root of the Google Chrome artifacts folder. In the example (Fig. 8.4—viewing Local State using Firefox JSONView add-on), we see that the three profiles are listed under "profile:info_cache:". We also see that the last active profile is profile 0: "Default". This can be important to you because by default Chrome will launch the last profile that was used. This end user experience can be configured differently but that's outside the scope of what we'll cover in this chapter.

Examining these profiles in the "Local State" file will provide you with valuable information. For example, you can establish when the profile was last launched by looking at the Epoch date value associated to the key "active_time" (Fig. 8.5).

You can tell if the user of that profile is currently logged into a Gmail account. There will be the below key:value pairs in the JSON file "Local State" that will only exist if the user is currently logged into a Gmail account. Note that others may also be unique to a logged in user, but these are confirmed to be associated to a logged in user based on the author's testing. These are found under "profile:info_cache:{profile}:" where {profile} is Default:, Profile 1:, Profile 2:, etc.

gaia_given_name: "name"

gaia_name: "full name"

gaia_picture_file_name: "name of profile picture file"

last_downloaded_gaia_picture_url_with_size: "Publicly accessible URL of their Google Account profile picture"

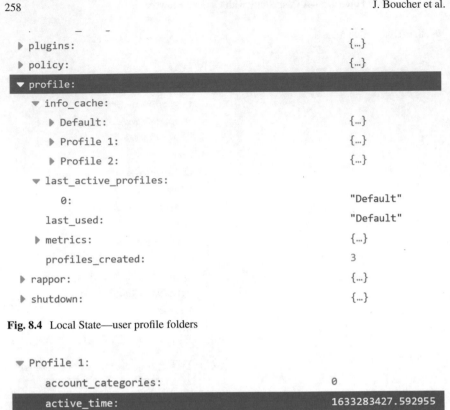

```
▶ plugins:                                                {...}
▶ policy:                                                 {...}
▼ profile:
   ▼ info_cache:
      ▶ Default:                                          {...}
      ▶ Profile 1:                                        {...}
      ▶ Profile 2:                                        {...}
   ▼ last_active_profiles:
         0:                                               "Default"
      last_used:                                          "Default"
   ▶ metrics:                                             {...}
      profiles_created:                                   3
▶ rappor:                                                 {...}
▶ shutdown:                                               {...}
```

Fig. 8.4 Local State—user profile folders

```
▼ Profile 1:
      account_categories:                                 0
      active_time:                              1633283427.592955
```

Fig. 8.5 Last time a profile was used

If a Google Chrome profile was used to log into a Gmail account and the user logged back out and stopped synching their data, some of the above key:value pairs will be present, but most will have a blank value.

If a Google Chrome profile never logged into a Google account, those key:value pairs will not be present under their profile in the Local State file.

You can also see what the last active profile was by examining "profile:info_cache:last_used:" (Fig. 8.6). If there is no value for that key, it means the user had the option selected to show the profile manager start-up window where you select which profile to use each time you launch Google Chrome. But the default behaviour is to simply launch Google Chrome from the last profile that was used when launching Google Chrome without any option to pick a specific profile.

```
last_active_profiles:                                     []
last_used:                                          "Profile 2"
```

Fig. 8.6 Last active profile

Google Chrome: 94.0.4606.71 (Official Build) (64-bit) (cohort: Stable)
Revision: 1d32b169326531e600d836bd395efc1b53d0f6ef-refs/branch-
heads/4606@{#1256}
OS: Windows 10 Version 20H2 (Build 19042.1237)
JavaScript: V8 9.4.146.18
User Agent: Mozilla/5.0 (Windows NT 10.0; Win64; x64) AppleWebKit/537.36
(KHTML, like Gecko) Chrome/94.0.4606.71 Safari/537.36
Command Line: "C:\Program Files
(x86)\Google\Chrome\Application\chrome.exe" --flag-switches-
begin --flag-switches-end
Executable Path: C:\Program Files (x86)\Google\Chrome\Application\chrome.exe
Profile Path: C:\Users\Jacques\AppData\Local\Google\Chrome\User
Data\Default

Fig. 8.7   chrome://version

On a live system with Google Chrome running (test system or system being seized), you can check the path to the profile folder by typing chrome://version in the URL and look at the "Profile Path:" (Fig. 8.7). On this page you will also see the Chrome version, OS information, and the command line options.

Local Versus Network Time

In addition to the above valuable insight you can extract from the Local State JSON file, you can also establish if the clock on the computer was accurate by examining network_time:network_time_mapping:local: versus network_time:network_time_mapping:network. In the example (Fig. 8.8) we see that the computer clock was off by 48 ms versus network time. Of course, this doesn't tell us if the clock was accurate an hour earlier, or sometime later when a user was still using the OS after exiting Google Chrome. But this gives you an additional data point you can examine when trying to ascertain the accuracy of a device's time.

There are many other potentially valuable key:value pairs in the Local State file. You are encouraged to explore this JSON file and do some of your own testing to confirm the significance of some of those values and how they might be of value to you in an examination. For example, what value might these key:value pairs with an Epoch timestamp tell you (Fig. 8.9)?

Or how about some of the values under "user_experience_metrics:"? The screenshot (Fig. 8.10) is but a few of the potentially useful values you'll find here. Of important note is that the timestamp "browser_last_live_timestamp" in this latest

```
▼ network_time:
   ▼ network_time_mapping:
      local:                                    1633284041958.114
      network:                                  1633284042000
```

Fig. 8.8   Local time versus network time

```
▼ uninstall_metrics:
      installation_date2:                                     "1562381879"
      launch_count:                                           "734"
      page_load_count:                                        "67449"
```

Fig. 8.9 Usage metrics

example is not expressed in Epoch time [14], but rather in Chrome WebKit [15] time. Whereas stats_buildtime is an Epoch date.

Or what might these values in Local State also with a Chrome WebKit timestamp tell you (Fig. 8.11)?

We know that Google tracks user activities. It's entirely plausible that Google may start tracking other user metrics in this JSON file in a future release of Google Chrome. Don't ignore this file in your analysis, as it could contains a few Easter eggs you likely were not aware of.

8.4 User Profile Preferences

In the previous section we saw that Local State is a JSON database file that contains settings that apply to all user profiles for that Google Chrome installation.

Within each profile folder (Fig. 8.3) found under the main Chrome user data folder (Table 8.1) you will find two other JSON database files of interest: "Preferences" and "Secure Preferences".

Within these two JSON files you will find various profile specific settings. Which files the settings are stored in will vary from OS to OS. So you may need to check both. Table 8.2 lists some potentially important key:value pairs and which file you will find them in depending on the OS.

Some key:value pairs have a default value. The absence of a key:value pair means the default applies as we see for popup blocking in Table 8.2. We see similar behaviour in Firefox preferences. Makes sense, why use up space storing a default value?

The key "account_info" has no value if that profile never logged into a Google account. If that profile did login at some point and logged back out, you will see the key with no value (Fig. 8.12).

Whereas if the profile is logged into their Google account, you will see a list of key:value pairs as you can see in Fig. 8.13. This is similar to what was observed in Local State. In some respects, this is redundant if you've already confirmed this via Local State. Or it can serve to validate what you observed in Local State.

In Local State we saw that it includes information about each of the profiles. Within the Preferences/Secure Preferences file this is also tracked. There are different key:value pairs that are affected depending on whether the profile ever signed into a Google account, if it's currently signed in, or if it's currently signed out (paused).

```
▼ user_experience_metrics:
      default_opt_in:                                    2
      low_entropy_source3:                               5023
      machine_id:                                        12974417
      pseudo_low_entropy_source:                         3825
      reporting_enabled:                                 false
      session_id:                                        767
   ▼ stability:
         browser_last_live_timestamp:                    "13277770546354724"
         child_process_crash_count:                      0
         crash_count:                                    0
         deferred_count:                                 0
         discard_count:                                  35
         exited_cleanly:                                 true
         extension_renderer_crash_count:                 0
         extension_renderer_failed_launch_count:         0
         extension_renderer_launch_count:                251
         gpu_crash_count:                                0
         incomplete_session_end_count:                   0
         launch_count:                                   5
         page_load_count:                                494
      ▶ plugin_stats2:                                   [...]
         renderer_crash_count:                           0
         renderer_failed_launch_count:                   0
         renderer_hang_count:                            0
         renderer_launch_count:                          153
         session_end_completed:                          true
         stats_buildtime:                                "1632361460"
         stats_version:                                  "94.0.4606.61-64"
```

Fig. 8.10 Other useful key:value pairs

```
▼ password_manager:
      os_password_blank:                           false
      os_password_last_changed:                    "13206838556248015"
```

Fig. 8.11 Other useful key:value pairs

Table 8.2 Preferences versus Secure Preferences

Pref	Default value	Windows 10	Mac OS	Ubuntu
account_info		Preferences	Preferences	Preferences
profile:		Preferences	Preferences	Preferences
session:restore_on_startup	5	Secure Preferences	Preferences	Preferences
session:startup_urls	<nil>	Secure Preferences	Preferences	Preferences
homepage:	<nil>	Secure Preferences	Preferences	Preferences
homepage_is_newtabpage	True	Secure Preferences	Preferences	Preferences
bookmark_bar:show_on…	False	Preferences	Preferences	Preferences
default_search_provider…	Google	Secure Preferences	Preferences	Preferences
profile:default_content_setting_values:popups:	Does not exist (block popups)	Preferences	Preferences	Preferences
profile:content_settings:exceptions:popups:	<nil>	Preferences	Preferences	Preferences

```
account_info:                                              []
```

Fig. 8.12 Never logged in

```
▼ account_info:
   ▼ 0:
        account_id:                     "106␣␣␣␣␣␣␣␣␣␣85"
        email:                          "jjrboucher@gmail.com"
        full_name:                      "Jacques Boucher"
        gaia:                           "10␣␣␣␣␣␣␣␣␣␣38985"
        given_name:                     "Jacques"
        hd:                             "NO_HOSTED_DOMAIN"
        is_child_account:               false
        is_under_advanced_protection:   false
        locale:                         "en"
      ▶ picture_url:                    "https://lh5.googleuserco_ru/g␣␣␣␣␣␣:-A/photo.jpg"
```

Fig. 8.13 Currently logged in

When you log out of a Google account in a Google Chrome profile, you still see that Google account atop of Google Chrome but it's showing as paused.

If a user logs into a new Google account from a profile already associated to another Google account, Google Chrome will ask the user if they want to create a

▼ account_info:

  ▶ 0:   {…}

  ▶ 1:   {…}

  account_tracker_service_last_update: "13278282522361843"

Fig. 8.14 Other Google accounts

new Google Chrome profile for that Google account (or switch to an existing one if one already exists). If the user accepts, a new Google Chrome profile is created for that newly signed in user (or it switches to the existing profile for that user). If the user declines, the user remains in that Google Chrome profile associated to the first Google account and any browsing activity under that profile continues to be associated to the Google account first associated to that profile.

There are three keys in the JSON file that the author noted are affected depending on if someone is logged in to a Google Chrome profile: "account_info:", "gaia_cookie:", and "profile:"

During testing with Google Chrome version 94, the author noted that the key:value pairs under "account_info:0:" were populated with details relating to the Google account associated to that Google Chrome profile (e.g., "email:", "full_name:", "given_name:", "picture_url"). The sub-key "account_info:account_tracker_service_last_update:" contains a UTC WebKit timestamp of when that Google Chrome profile was created on the device in question.

If other Google accounts are also logged into under the same profile, you will see those listed under the sub-key "account_info:1:", "account_info:2:", etc. (Fig. 8.14).

The key "profile:" also contains potentially relevant information. It contains the version of Google Chrome that created that profile ("created_by_version") on that device, as well as a WebKit UTC timestamp of when the profile was created on the device ("creation_time").

The subkey "profile:last_engagement_time:" initially appeared to capture the last time that Google Chrome profile was shut down. During the author's testing, however, in one testing scenario that timestamp did not correspond to that, rather it was from almost two hours earlier.

This illustrates the challenge of forming a definitive conclusion based on the key:value pairs in these JSON database files related to Google Chrome. With Google Chrome now on a 4-week rapid release cycle, a user who has been using an install of Google Chrome over a period of 12 months will have 26 updates during this time. Any one of these updates could add, remove, or change some of the key:value pairs in these JSON files.

These JSON files are a rich source of artifacts that could contain the missing piece of the puzzle. But you need to be cautious when stating an opinion based on these artifacts. As with everything else in digital forensics, validate/corroborate your findings.

8.5   Analyzing SQLite Files

8.5.1   Secure_Delete

SQLite can be compiled with many different options enabled or disabled via Pragma statements [16]. One of those is a Pragma statement called secure_delete [17]. If this option is enabled by the developer of the application, any record deleted in the SQLite file will be immediately overwritten with the hexadecimal value 00.

Enabling secure_delete ensures greater privacy, but it does come at a cost of performance. Firefox has implemented SQLite with secure_delete enabled. Whereas Google Chrome opted for performance and implemented SQLite without secure_delete.

When secure_delete is not enabled, the record is simply marked as deleted and eventually either overwritten by a new record, or the database is compressed via the vacuum command [18]. Because Google Chrome did not enable secure_delete, it instead periodically runs the vacuum command to clean up and compact the database file. This means that you could get a keyword hit on a Google Chrome SQLite file but when you examine its content with a regular SQLite tool, you will not see the transaction. That will let you know that you hit on a deleted, but not yet overwritten record.

8.5.2   Analysis Tools

There are a few commercial tools that can help you analyze an SQLite file in addition to general forensic as well as browser analysis tools able to parse Google Chrome browsing data. To avoid the risk of endorsing a commercial tool over another, the author will only reference a free, open source multi platform SQLite tool called DB Browser for SQLite [19]. This free tool allows you to open an SQLite file as read/write, or read-only, with the latter being the recommended option when conducting your analysis.

DB Browser for SQLite will not analyze journal files. If they are present and you open the database read-write, the journal transactions will be processed. DB Browser for SQLite will also not find deleted records in an SQLite file.

If you are conducting forensics on a limited budget, you can try opening the SQLite file without the journal file, and in a second instance open it with the journal file and analyze both. This is not without risk. For example, you will not see a new transaction still in the journal file that has a deletion also pending in the journal file.

As for deleted transactions, some of the commercial tools that parse SQLite journal files also search for deleted records. If you do not have such a tool, but have a main forensic tool, you will still get keyword hits on deleted records. But if you don't search for the right keyword, you won't know about those deleted records.

Database Is Locked

If you are doing some testing and attempt to open a Google Chrome database from a Google Chrome profile which is running, it will fail as the database is locked. If you are using DB Browser for SQLite, you will get the error.

You will need to close the Google Chrome instance associated to that profile and try again. If you continue to get the error after exiting all instances of Google Chrome, it's likely that a Google Chrome process is still running. You will need to display your running processes and delete that process.

8.6   History and Typed URLs

As noted earlier in this chapter, Google Chrome allows you to create multiple user profiles, each containing the user data for that profile (e.g., user preferences, browser history, form data, saved passwords, etc.). Table 8.1 in this chapter provides you with the path to the root of Google Chrome's user folders. From there, you will see the default profile in the folder called "Default" (i.e., %localapp-data%/Google/Chrome/User Data/{Chrome Profile}/—Default in this example). If other profiles exist, they will be in folders called "Profile 1", "Profile 2", etc. Thus, any reference to an artifact in this section will be contained within the profile folder.

Google Chrome's history is stored in an SQLite file called "history" with no extension, and an associated journal file called "history-journal". Google Chrome implements SQLite using a roll back journal. This is used to recover if the database is not shut down cleanly. If the journal file is 0 kB, you don't have to worry about any pending transactions in it requiring analysis. If there is data in the journal file, you will want to use a tool that supports analyzing the journal file to examine any pending transactions.

Google Chrome retains the last 90 days of browsing history. Artifacts related to browsing (e.g., urls visited, cookies, cache) are all subject to this 90-day limit. Whereas other artifacts such as preferences, form data, and passwords are kept until the user opts to delete them.

Data in Google Chrome extensions not subject to this 90-day limit. Each extension developer decides what artifacts they track, and how long it is retained, as this is stored in the extension's sub-folder.

One approach to analysis of Google Chrome artifacts can involve copying a Google Chrome profile folder out of your forensic tool and pasting those files into a Google Chrome profile folder you created on your forensic machine or in a VM and then deleted the original content, retaining only the root folder. This approach allows you to launch Google Chrome (either with or without connectivity depending on your objective) and view the artifacts from within Google Chrome itself (e.g., view the browser's history, cookies, preferences, etc.).

An important caveat if using this approach is that any history artifacts older than 90 days will be purged, so you will lose some content using this approach unless you back date your system's date. The author's experience when testing this a few years ago was that Google Chrome does not wipe all content older than 90 days immediately on launch. The deletion took place over time (exact time not tested), as shutting down the browser after a minute, or a few minutes revealed that some of the old artifacts had been purged, but not all of them. With each re-launch, more artifacts were deleted.

8.6.1 SQLite Tables of Interest

At the time of this writing, there are 17 tables in the SQLite file "history". We will not explore all of them, but rather the main ones relating to browsing history. Chances are your forensic tool also focuses only on the main ones, and likely not all the fields within those tables. Attempting to support every artifact in the tables, validating for each new release coming out every 4 weeks, including adding new artifacts as they appear and deleting old ones as they are become obsolete would be far too labour intensive. The value that you would yield from that would not be proportional to the effort required to maintain this level of support.

In this section we will focus on three tables: "urls", "visits", and "visit_source".

"urls" Table

The "urls" table has seven fields at the time of this writing (Table 8.3).

When navigating pages in Google Chrome, each complete URL is stored in this table. If the URL is already in the table, it is not stored a second time. Rather the fields "visit_count", "typed_count", and "last_visit_time" are updated as applicable. When a URL goes beyond the 90 days and its record is removed, the "visit_count" is decreased accordingly. In other words, the "visit_count" is the total # of visits in

Table 8.3 URLS table fields

Field	Description
id	Unique id that auto-increments
url	Unique URL
title	Associated page title
visit_count	Number of times the URL was visited in the past 90 days
typed_count	Number of times the URL was typed
last_visit_time	Timestamp when the URL was last visited (Google WebKit UTC time)
Hidden	If a URL is hidden, it is not included in the autocomplete [20]

id		url	title	visit_count	typed_count	last_visit_time	hidden
Filter	Filter		Filter	Filter	Filter	Filter	Filter
459	15826	https://www.canada.ca/en/services/jobs/...	Government of Canad...	4	0	13274137247463631	0
460	15827	https://www.telus.com/en/	Phones, Internet and ...	1	0	13274652238173967	0
461	15828	https://mail.google.com/mail/u/0/x/...	https://...	2	0	13273960656365998	0
462	15830	http://lexus.ca/	Lexus Canada: Luxury ...	3	3	13274499222716016	0
463	15832	https://m.youtube.com/	Home - YouTube	7	0	13273725899249986	0
464	15833	https://webmail.bell.net/bell/	https://webmail.bell.n...	2	0	13273958527072393	0
465	15834	https://www.homedepot.com/b/Tools-Air-...	https://...	6	1	13274242895215941	0
466	15835	http://apple.ca/	Apple (Canada)	1	1	13274921670040390	0
467	15836	https://who.is/	WHOIS Search, Domai...	2	2	13274028771572368	0
468	15837	https://covertinstruments.com/	Covert Instruments	2	1	13274406725854325	0

Fig. 8.15 "urls" table

the browser's history. Because it's only kept for 90 days, it's the total visit count for the last 90 days.

Let's look at what you could conclude if you only analysed this table by looking at sample from it (Fig. 8.15).

We can see the URLs that were visited, the page title for each URL, how many times it was visited, how many times it was typed, and the last time it was visited. Let's look at row #465. The record # is 15834, the URL is https://www.homedepot. com/b/Tools-Air-Compressor-Tools-Nail-Guns/N-5yc1vZc2cd/Ntk-EnrichedProd uctInfo/Ntt-nailer?Ntx=mode+matchpartialmax&NCNI-5&visNavSearch=nailer, the page title is https://www.homedepot.com/b/Tools-Air-Compressor-Tools-Nail-Guns/N-5yc1vZc2cd/Ntk-EnrichedProductInfo/Ntt-nailer?Ntx=mode+matchpart ialmax&NCNI-5&visNavSearch=nailer, the visit count is 6, it was typed 1 time, and the last time it was visited was at 13274242895215941 Google WebKit time (UTC).

There are different ways to convert WebKit and epoch timestamps. The author has found https://www.epochconverter.com/WebKit to work very well when converting a single value. The WebKit time in the previous paragraph converts to Tuesday, August 24, 2021, 1:41:35 AM UTC.

But if you wanted to see the converted WebKit time for all records in this table, you could do that with the following SQLite query:

```
SELECT    id,
          url,
          title,
          visit_count,
          typed_count,
          last_visit_time,
          datetime(last_visit_time/1000000-11644473600,'unixepoch')
AS "Decoded last_visit_time"
FROM      urls
```

Note that we are using the SQLite datetime() function to convert the WebKit time to 'unixepoch' which the function can then convert to human readable format.

Best Practice Recommendations

- When decoding a field in your SQLite statement, first display the raw field, followed by the decoded value. You want to avoid being on the stand and unable to testify to the raw values in the database, only your decoded values. You want the raw value that was in the database alongside your decoded value.
- Make generous use of comments in your SQLite statements so that you will understand what they do 6, 12, or 18 months later when having to explain it to a judge or jury, and include a copy of your SQLite queries in your report showing how you produced the output.

Using the same urls table as earlier, here is what the same records would look like (Fig. 8.16).

This is certainly more convenient than manually decoding the WebKit times using a website. But there is a limitation to the SQLite statement we just used. Some URLs could have a last_visit_time of 0. For those, the converted time will end up being 1601-01-01 00:00:00. Rather than displaying an inaccurate date, you can modify the SQLite statement to deal with that by changing the datetime() statement to the following:

	id	url	title	visit_count	typed_count	last_visit_time	Decoded last_visit_time
459	15826	https://www.canada.ca/en/...	Government of Canada jobs -...	4	0	13274137247463631	2021-08-22 20:20:47
460	15827	https://www.telus.com/en/	Phones, Internet and TV on ...	1	0	13274652238173967	2021-08-28 19:23:58
461	15828	https://mail.google.com/mail/u/...	https://mail.google.com/mail...	2	0	13273960656365998	2021-08-20 19:17:36
462	15830	http://lexus.ca/	Lexus Canada: Luxury Sedan,...	3	3	13274499222716016	2021-08-27 00:53:42
463	15832	https://m.youtube.com/	Home - YouTube	7	0	13273725899249986	2021-08-18 02:04:59
464	15833	https://webmail.bell.net/bell/	https://webmail.bell.net/bell/...	2	0	13273958527072393	2021-08-20 18:42:07
465	15834	https://www.homedepot.com/b/...	https://www.homedepot.com...	6	1	13274242895215941	2021-08-24 01:41:35
466	15835	http://apple.ca/	Apple (Canada)	1	1	13274921670040390	2021-08-31 22:14:30
467	15836	https://who.is/	WHOIS Search, Domain Nam...	2	2	13274028771572368	2021-08-21 14:12:51

Fig. 8.16 "urls" table, decoded WebKit time

```
CASE
    WHEN last_visit_time IS 0 THEN NULL
    ELSE datetime(last_visit_time/1000000-11644473600,'unixepoch')
END AS "Decoded last_visit_time"
```

If all you are interested in is knowing is how often someone visited a URL, how many times they typed it, and when they last visited it, this works. But URL #15834 for homedepot.com was visited 6 times and typed 1 time. We only know the last time it was visited, not the other 5 times, and we don't know which time it was typed.

"visits" Table

The "visits" table (Table 8.4) has nine fields at the time of this writing, but we will focus on the six.

This table tracks each visit to a URL. If you visit the exact same URL 5 times, there will be 5 records in this table all pointing to a single record in the "urls" table we covered previously. Here is a raw view of a sampling of records from the "visits" table (Fig. 8.17).

This table also contains a date value expressed in Chrome WebKit UTC timestamp which will need to be converted. The transition value is another field that will need to be decoded which we'll tackle a bit further in this chapter. The "url" field is what is known as a foreign key. It points to a record in another table, the "urls" table in this case.

Table 8.4 VISITS table fields

Field	Description
id	Unique id that auto-increments
url	Foreign key that points to the id of the URL in the urls table
visit_time	Timestamp when the URL was visited (Google WebKit UTC time)
from_visit	Points to the id in the "visits" table from where the user navigated from prior this URL
transition [21]	Transition value that tracks several qualifiers about the visit
visit_duration	Visit duration in milliseconds

	id	url	visit_time ▾¹	from_visit	transition	segment_id	visit_duration	incremented_omnibox_typed_score	publicly_routabl
	Filter	Filter	Filter	Filter	Filter	Filter	Filter	Filter	Filter
1	45956	15987	13270602386975707	0	805306368	0	0		0
2	46535	173	13270604956567473	0	1610612736	0	0		0
3	46536	173	13270604959347996	0	1610612736	0	0		0
4	45090	15764	13270605015587216	0	-2147483647	0	0		0
5	45221	15775	13270605015587216	0	268435457	0	0		1
6	46537	173	13270605279977646	0	1610612736	0	0		0
7	46538	173	13270605505863838	0	1610612736	0	0		0
8	44888	176	13270610932800826	0	301989889	0	0		1
9	44889	176	13270610933346649	0	-2147483648	0	0		0
10	44890	176	13270611813341930	0	-1610612729	0	0		0
11	44891	176	13270611813266672	0	1610612729	0	0		0

Fig. 8.17 "visits" table

270 J. Boucher et al.

When we looked at the "urls" table earlier, we looked at a record relating to a visit to a page on homedepot.com. The id of that record was 15834 (Fig. 8.18). It was visited 6 times, of which one of those was a typed URL. And we decoded the last visit time as Tuesday, August 24, 2021, 1:41:35 AM UTC.

The following SQLite statement was used to focus on the "visits" entries for url 15834.

```
SELECT id, url, from_visit, visit_time,
CASE /*if visit_time is NULL don't convert, leave it NULL.
    WHEN visit_time is NULL THEN NULL
    ELSE datetime(visit_time/1000000-11644473600,'unixepoch')
END AS 'Decoded visit_time (UTC)',
transition, visit_duration, visit_duration/1000000 AS "Visit Duration in
seconds"
FROM visits
WHERE url=15834
```

That query produced the results (Fig. 8.19).

Now we see the timestamp for all six visits, with the last one matching the timestamp decoded earlier in the "urls" table relating to this record (Fig. 8.16). The "url" value is an id value that points to a record in the "urls" table.

If all you must link is a handful of records from the "visits" table to their URL in the "urls" table, doing this manually is possible. But it's not feasible to do this for all records in a user's 90-day browsing history. We can use a JOIN command in SQLite to link two tables together.

The following SQLite statement omits a few fields for brevity of the demonstration.

	id	url	title	visit_count	typed_count	last_visit_time	hidden
1	15834	https://www.homedepot.com/b/Tools-Air-...	https://...	6	1	13274242895215941	0

Fig. 8.18 "urls" table, record 15834

	id	url	from_visit	visit_time	Decoded visit_time (UTC)	transition	visit_duration	Visit Duration in seconds
1	45408	15834	0	13273704915958255	2021-08-17 20:15:15	805306368	0	0
2	45409	15834	0	13273704916385598	2021-08-17 20:15:16	805306368	0	0
3	45410	15834	0	13273704916385598	2021-08-17 20:15:16	805306368	0	0
4	45411	15834	0	13273704917280179	2021-08-17 20:15:17	805306368	0	0
5	45412	15834	0	13273704917289074	2021-08-17 20:15:17	805306368	0	0
6	45413	15834	0	13274242895215941	2021-08-24 01:41:35	838860801	0	0

Fig. 8.19 Querying URL 15834 in the "visits" table

> SELECT visits.id, urls.url, visits.from_visit, visits.visit_time,
> CASE
>   WHEN visits.visit_time is NULL THEN NULL
>   ELSE datetime(visits.visit_time/1000000-11644473600,
> 'unixepoch')
> END AS 'Decoded visit_time (UTC)',
> visits.transition, urls.typed_count, visits.visit_duration AS "visit duration
> in milliseconds"
> FROM visits
> LEFT JOIN urls ON urls.id = visits.url

The output of that statement is as in Fig. 8.20.

This gives us a pretty good view of the browsing activity of the user. If we filter on record 15834, we have Fig. 8.21.

The doesn't tell us which URL was typed. We only know that one of the six was typed, but not which one. One might be inclined to conclude that record #6 (Fig. 8.21) is the typed one, as its transition value is different than the other five. In this case we are lucky that all the other transition values are identical so we can take an educated guess without decoding it. But in other cases it won't be that obvious. Plus, you will miss other valuable info if you do not decode the value.

	id	url	from_visit	visit_time	Decoded visit_time (UTC	transition	typed_count	sit duration in milliseconi
77	44357	https://www.lexus.com/	0	13271624486058408	2021-07-24 18:21:26	838860801	19	0
78	44358	https://www.lexus.com/	0	13271626576389310	2021-07-24 18:56:16	838860801	19	0
79	44713	https://mail.google.com...	0	13271628695046115	2021-07-24 19:31:35	838860801	24	0
80	44714	https://mail.google.com...	0	13271628699338462	2021-07-24 19:31:39	268435456	24	0
81	44715	https://mail.google.com...	0	13271629430784320	2021-07-24 19:43:50	1610612736	24	0
82	44425	https://...	0	13271637922889046	2021-07-24 22:05:22	838860801	6	0
83	44359	https://www.lexus.com/	0	13271638038454183	2021-07-24 22:07:18	838860801	19	0
84	45619	https://www.toyota.com/	0	13271638044707185	2021-07-24 22:07:24	838860801	6	0
85	45620	https://www.toyota.com/	0	13271638044707185	2021-07-24 22:07:24	838860801	6	0
86	44379	https://www.bmw.com/...	0	13271638313283840	2021-07-24 22:11:53	268435457	7	0
87	44380	https://www.bmw.com/...	0	13271638313283840	2021-07-24 22:11:53	268435457	7	0

Fig. 8.20 Joining "urls" and "visits" tables

id	url	from_visit	visit_time	Decoded visit_time (UTC	transition	typed_count	sit duration in millisecon
1	45408	https://www.homedepot.com/...	0	13273704915958255	2021-08-17 20:15:15	805306368	1	0
2	45409	https://www.homedepot.com/...	0	13273704916385598	2021-08-17 20:15:16	805306368	1	0
3	45410	https://www.homedepot.com/...	0	13273704916385598	2021-08-17 20:15:16	805306368	1	0
4	45411	https://www.homedepot.com/...	0	13273704917280179	2021-08-17 20:15:17	805306368	1	0
5	45412	https://www.homedepot.com/...	0	13273704917289074	2021-08-17 20:15:17	805306368	1	0
6	45413	https://www.homedepot.com/...	0	13274242895215941	2021-08-24 01:41:35	838860801	1	0

Fig. 8.21 Joining "urls" and "visits" tables, record 15834 from "urls" table

The transition value actually tells us a lot about a URL as we can see from Chromium's source code here: https://source.chromium.org/chromium/chromium/src/+/master:ui/base/page_transition_types.h.

The transition value is a four-byte value. Decoding the value requires us to do some bitwise operations. The right most byte tells us the page transition. The three left most bytes are qualifiers for that transition.

Let's look at record #1 in the example (Fig. 8.21) which has a transition value of 805306368. First, we must convert the value to binary which gives us the following value: 00110000 00000000 00000000 00000000.

The right-most byte has a value of decimal 0. Lines 28–31 of the earlier referenced source code has the following:

```
PAGE_TRANSITION_FIRST = 0,
// User got to this page by clicking a link on another page.
PAGE_TRANSITION_LINK = PAGE_TRANSITION_FIRST,
```

So, we know that the user navigated to this URL by clicking on a link.

Here is Table 8.5 of the ten possible values according to Chromium's source code.

Next, we need to do some bitwise operations on the remaining three bytes to determine which qualifier(s) are applicable to this URL visit.

Digging further into the source code for page_transition_types.h [21], we find the possible qualifiers (Table 8.6) and which bit(s) must be on to be applicable.

As a reminder, the binary value of the transition we are decoding is: 00110000 00000000 00000000 00000000 where bits 3 and 4 are on.

Bit 4 corresponds to the qualifier PAGE_TRANSITION_CHAIN_START, and Bit 3 corresponds to the qualifier PAGE_TRANSITION_CHAIN_END. The source code further describes the chain end as "The last transition in a redirect chain". That

Table 8.5 Page transition

Page transition description	Value
PAGE_TRANSITION_LINK	0
PAGE_TRANSITION_TYPED	1
PAGE_TRANSITION_AUTO_BOOKMARK	2
PAGE_TRANSITION_AUTO_SUBFRAME	3
PAGE_TRANSITION_MANUAL_SUBFRAME	4
PAGE_TRANSITION_GENERATED	5
PAGE_TRANSITION_AUTO_TOPLEVEL	6
PAGE_TRANSITION_FORM_SUBMIT	7
PAGE_TRANSITION_RELOAD	8
PAGE_TRANSITION_KEYWORD	9
PAGE_TRANSITION_KEYWORD_GENERATED	10

Table 8.6 Page qualifiers

Page transition qualifier	Hex value	Binary value
PAGE_TRANSITION_BLOCKED	0x00800000	0b0000 0000 1000 0000 0000 0000
PAGE_TRANSITION_FORWARD_BACK	0x01000000	0b0000 0001 0000 0000 0000 0000
PAGE_TRANSITION_FROM_ADDRESS_BAR	0x02000000	0b0000 0010 0000 0000 0000 0000
PAGE_TRANSITION_HOME_PAGE	0x04000000	0b0000 0100 0000 0000 0000 0000
PAGE_TRANSITION_FROM_API	0x08000000	0b0000 1000 0000 0000 0000 0000
PAGE_TRANSITION_CHAIN_START	0x10000000	0b0001 0000 0000 0000 0000 0000
PAGE_TRANSITION_CHAIN_END	0x20000000	0b0010 0000 0000 0000 0000 0000
PAGE_TRANSITION_CLIENT_REDIRECT	0x40000000	0b0100 0000 0000 0000 0000 0000
PAGE_TRANSITION_SERVER_REDIRECT	0x80000000	0b1000 0000 0000 0000 0000 0000
PAGE_TRANSITION_IS_REDIRECT_MASK	0xC0000000	0b1100 0000 0000 0000 0000 0000

suggests that this is a redirect URL arising from the initial URL that a user navigated to. The fact that a URL in "visits" is both the start and end of the chain means it's the only URL in that chain.

Now let's decode the transition value for the last record in this series of six visits, which has a value of 838860801 (Fig. 8.21). The binary representation is:

00110010 00000000 00000000 00000001

The right-most byte in this case has a value of decimal 1, which means it's a typed URL as we suspected (but it won't always be that obvious). The three left most bytes are then used to determine the applicable qualifiers.

Bits 3 and 4 from the left are respectively the end and start of the page transition chain. The seventh bit from the left is PAGE_TRANSITION_FROM _ADDRESS_BAR. Meaning this URL was navigated to via the address bar. Makes sense since it's a typed URL.

Putting all this together, the visits record with id 45413 that we examined earlier is a typed URL, and its qualifiers are that it's from the address bar and is both the start and end of the navigation chain.

Fortunately for us, SQLite supports bitwise operations, so we can parse the transition value within an SQLite statement. For example, to parse the right most byte, you can use the following SQLite CASE statement to perform a bitwise operation on the transition value and isolate the bits in the right most byte:

```
CASE (transition&0xff)
        WHEN 0 THEN 'Clicked on a link'
WHEN 1 THEN 'Typed URL'
WHEN 2 THEN 'Clicked on suggestion in the UI'
WHEN 3 THEN 'Auto subframe navigation'
WHEN 4 THEN 'User manual subframe navigation'
WHEN 5 THEN 'User typed text in URL bar, then selected an entry that
did not look like a URL'
WHEN 6 THEN 'Top level navigation'
WHEN 7 THEN 'User submitted form data'
WHEN 8 THEN 'User reloaded page (either hitting ENTER in address bar,
or hitting reload button)'
WHEN 9 THEN 'URL generated from a replaceable keyword other than
default search provider'
WHEN 10 THEN 'Corresponds to a visit generated for a keyword.'
ELSE 'New value!: '||transition&0xff||' Check source code for meaning!'
END AS 'Transition Type',
```

Best Practice Recommendation

We know that a single byte can have values between 0 and 255 (000000–111111), thus 256 values. But here Google Chrome is currently only using values 0–10. When writing SQLite statements like this where only 11 of the possible 256 values are being used, you can future proof your statement by including the ELSE clause as above. If a value other than 0–10 is found, it will result in "New value!: {value}" being displayed for that field where {value} is the new value it found. This alerts you to the fact that the value is something other than one of the values you were expecting. Seeing this you can research the new value and update your SQLite statement accordingly.

The bitwise operation to decode the bits in the other three bytes is done each with its own CASE statement. It would be too lengthy to include all of them in this chapter, but here is an example of one of those case statements that checks to see if the bit is on denoting that the user navigated to the URL via the forward/backward button:

```
CASE (transition&0x01000000) /* Applies mask to isolate 25th bit from
the right */
        WHEN 0x01000000 THEN 'yes' /*bit is set */
END AS 'Navigated using Forward/Back button',
```

And here is an example of the CASE statement to check if the URL was the result of a redirect sent from the server by HTTP headers.

```
CASE (transition&0x80000000) /* Applies mask to isolate 32nd bit from
the right */
        WHEN 0x80000000 THEN 'yes' /* bit is set */
END AS 'Redirects sent from the server by HTTP headers.'
```

What Is a Typed URL?

We saw earlier that by decoding the transition value, we can determine if a specific visit was typed or not. But what is a typed URL? Most would answer that it's when you type a URL in the address bar (or omnibox as Google Chrome calls it), and that's correct. But are there other interactions that you might not have considered that would result in a "typed" URL?

Understanding how to decode the records in an SQLite file is important. But equally important is understanding what user actions will cause that data to be written to the database you are analyzing. Does a value in a record denote a very specific user action, or are their different ways a user can interact with the application to cause the same results to be written to the database? It's the difference between stating that because you retrieved a particular value from a record, it means a user did A, versus a user could have done A, B, or C to cause that value to be written to the database.

The user copies a URL and pastes it into the address bar and hits <ENTER>. Will that result in a typed URL?

The user enters text in the address bar that results in a search with the browser's default search engine. Will that result in a typed URL?

The user edits the URL in the address bar. For example, they search for "cats and hogs" but meant to search for "cats and dogs". Instead of editing the text in the Google search box on Google's search page, they go into the address bar and changes hogs to dogs and hit <ENTER>. Will that result in a typed URL?

The user simply clicks in the address bar, makes no changes to the URL and hits <ENTER>. Will that result in a typed URL?

The user types a URL, but it results in a redirect (e.g.: typing www.google.com will initially go to http://www.google.com, and then redirect to https://www.google.com). Will both records result in a typed URL?

The user types an invalid URL. Will that result in a Typed URL?

The user types a valid URL but the site is down. Will that result in a Typed URL?

The user types a valid URL but the device was not connected to the Internet at the time. Will that result in a typed URL?

You are encouraged to test this with not only Google Chrome, but any other browser where you find the smoking gun in a typed URL. You need to know what that browser considers a typed URL, and how it stores that data in its history file. Imagine trying to suggest that the user typed:

https://www.canadapost-postescanada.ca/cpc/en/support.page?ecid=murl_ddn_jb_100#panel2-5

That is a typed URL in the author's Google Chrome browser history. Clearly, the author did not type that. How would your credibility be impacted if you got on the stand and claimed that the user typed that entire string in the address bar which is why it's showing up as a typed URL in the database you analyzed?

"visit_source" Table

The "visit_source" table has two fields at the time of this writing (Table 8.7).

In the author's experience, the value 1 is not used in the field "value" in the "visit_source" table in Google Chrome. Rather Google Chrome simply does not store the id of the "visits" record if it's been browsed by the user. This is a common practice we observed when looking at the JSON files earlier. Why store all the URLs here as well as in the "visits" table. It's redundant. Hence why any URLs not in this table are locally browsed URLs, not synced URLs.

The most important value you will want to check is for a value of 0, indicating the URL was synced to this device after having been visited on another device. This may be important if you need to address the sync defence (it wasn't me, it synced from another device). Syncing only happens across browsers logged into the same Google account naturally.

In Fig. 8.22 we have an example of what you might see in the "visit_source" table.

The id is a foreign key that points to "visits.id". In the above example, it tells us that the "visits" record with ids 44357, 44358, 44359, 44379, 44380, and 44381 synced from another device.

Table 8.7 VISIT_SOURCE table fields	Field	Description
	id	Foreign key that points to visits.id

	source [22]	Value between 0 and 5
		SOURCE_SYNCED = 0, // Synchronized from somewhere else
		SOURCE_BROWSED = 1, // User browsed
		SOURCE_EXTENSION = 2, // Added by an extension
		SOURCE_FIREFOX_IMPORTED = 3,
		SOURCE_IE_IMPORTED = 4,
		SOURCE_SAFARI_IMPORTED = 5,

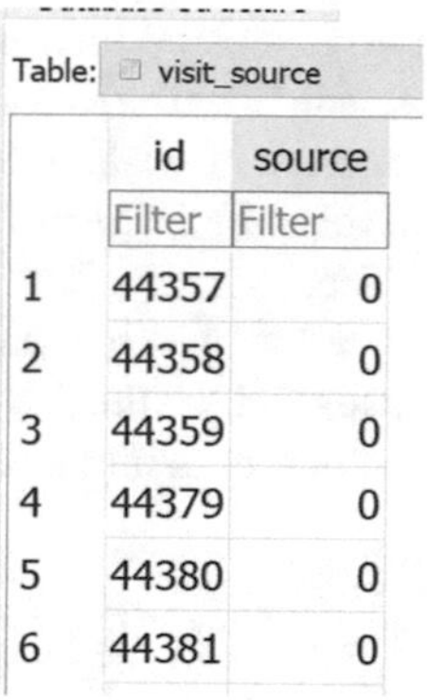

Fig. 8.22 Sample
"visit_source" table

	id	source
	Filter	Filter
1	44357	0
2	44358	0
3	44359	0
4	44379	0
5	44380	0
6	44381	0

Table: ▢ visit_source

As we saw with the "visits" and URLs tables, you can use the JOIN statement to query across tables, denoting how they are linked together.

8.7 Downloads

Google Chrome tracks downloads in the same SQLite file as it tracks history, but in different tables. The following tables found in the "history" SQLite file have names that suggest to us that they are used to track downloads: downloads, downloads_reroute_info, downloads_slices, downloads_url_chains.

The author has over 400 downloads in the "downloads" table. You could simply browse that table and yield some good insight in the downloads. The only other table with records in it was "downloads_url_chains". Accordingly, the other two tables will not be reviewed.

"downloads" Table

The "downloads" table has 26 fields at the time of this writing, but we will focus on the 13 (Table 8.8).

When parsing the interrupt_reason in your SQLite statement, you will want to adopt the earlier recommended best practice of alerting you if a value other than one of the expected values is encountered so that you can research the new value and update your SQLite statement.

"downloads_url_chains" Table

The "downloads_url_chains" table has three fields at the time of this writing (Table 8.9).

This table can also be manually examined by browsing it with your favorite SQLite browser. Figure 8.23 is a sample of what you might see in it.

Note that the id is a foreign key that points to the id in the "downloads" table. In the above example, id 1 and 2 each only have one URL. Next, we see id 4 with two chains. Which means it redirected from the first to the second to perform the download. If we look at id = 5, we see 5 entries in the chain. When examining the record with id #5 in the "downloads" table, you must also consider these five records in this table. The last record will contain the URL where the download actually took place.

We can write an SQLite statement that will return each record in "downloads_url_chains" along with the relevant data from "downloads". Unfortunately, the output is much too wide to be able to display on a page.

Alternate Data Stream (ADS)

When you download a file from the Internet on a Windows computer and save it to an NTFS file system, in addition to the file it will create an alternate data stream. We won't be covering ADS in this chapter. But it's worth reading up on it and being familiar with the ADS created alongside a file when it's downloaded from the Internet. This is not unique to Google Chrome.

Table 8.8 DOWNLOADS table fields

Field	Description [23]
id	Unique id that auto-increments
current_path	Current disk location
target_path	Final disk location
start_time	When the download was started. Google WebKit UTC timestamp
end_time	When the download completed. Google WebKit UTC timestamp
last_access_time	The last time it was accessed. Google WebKit UTC timestamp
last_modified referrer site_url tab_url tab_referrer_url state interrupt_reason	Last-modified header. Text UTC timestamp HTTP referrer Site URL for initiating site Tab URL for initiator Tag referrer URL for initiator 1 = complete, 4 = interrupted Download Interrupt Reason [24]. Current values 0–3, 5–7, 10–15, 20–24, 30–39, 40–41, 50

Table 8.9 DOWNLOADS_URL_CHAINS table fields

Field	Description [23]
id	Foreign key that points to downloads.id
chain_index	Index of url in chain. 0 is initial target, MAX is target after redirects
url	URL

Fig. 8.23 downloads_url_chains table

Different Ways to Download

There are a few different ways you can download a file from your browser. You can choose to download the file by clicking on the link. You can right click and choose Save As. Or you can choose to run the file. You should test to see what the downloads artifacts look like for each scenario to better understand what the artifacts represent.

8.8 Search Terms

Proving intent is one of the challenges a forensic examiner faces. A user may claim that they did not intentionally navigate to a particular URL. There is a table in the "history" SQLite file called "keyword_search_terms".

The "keyword_search_terms" table in the "history" SQLite file has four fields at the time of this writing (Table 8.10).

You can see that to make sense of this information you not only have a foreign key pointing to another table within the same SQLite file (history), but you also have a foreign key pointing to a table in table in another SQLite file (web data). Manually parsing this would be tedious. Before we tackle that, let's look at an example of what's in the "keywords_search_term" table in the "history" SQLite file (Fig. 8.24).

We see a lot of entries that point to record id = 2 in the "keywords" table in "web data" SQLite file. How do we know this from looking at Fig. 8.24? The first field

Table 8.10 KEYWORD_SEARCH_TERMS table fields

Field	Description
keyword_id	Foreign key that points to the field "id" in the "web data" SQLite file, "keywords" table
url_id	Foreign key that points to urls.id in the "history" SQLite file
term normalized_term	The search term typed by the user The search term converted to all lower case

	keyword_id	url_id	term	normalized_term
	Filter	Filter	Filter	Filter
35	2	16143	kodak 110 film	kodak 110 film
36	2	16147	kodak 110 film ...	kodak 110 film developing new jersey
37	2	16148	kodak 110 film ...	kodak 110 film developing new jersey
38	2	16149	kodak 110 film ...	kodak 110 film developing new jersey
39	2	16174	"chain of custody" ...	"chain of custody" guide digital evidence
40	2	16176	"chain of custody" ...	"chain of custody" guide digital evidence jacob
41	2	16187	MS Word missing ...	ms word missing insert caption to picture
42	2	16208	db browser	db browser
43	2	16209	sqlite secure delete	sqlite secure delete
44	2	16217	sqlite secure delete	sqlite secure delete
45	50	16231	halloween makeup ...	halloween makeup tutorial
46	153	16231	halloween makeup ...	halloween makeup tutorial
47	50	16232	sqlite android tutorial	sqlite android tutorial
48	153	16232	sqlite android tutorial	sqlite android tutorial
49	166	16235	Garden State Parkway	garden state parkway
50	166	16237	Empire State Building	empire state building
51	166	16238	Empire State Building	empire state building
52	82	16246	Ford Mustang Mach E	ford mustang mach e

Fig. 8.24 keywords_search_terms table

is called "keyword_id". Through testing, the author determined that it points to the "id" column in the "keywords" table in the Google Chrome SQLite file called "web data".

The "keywords" table (Table 8.11) in the "web data" SQLite file has 24 fields at the time of this writing. We will only use 5 of them.

Here is a basic query of the keywords table, extracting the fields above and producing the output in Fig. 8.25.

Table 8.11 KEYWORDS table fields

Field	Description [25]
id	Unique id that auto-increments
short_name	The description of the search engine
keyword	The search engine keyword for omnibox access
url	The actual parameterized search engine query URL
date_created	The date this search engine entry was created (WebKit UTC time)

	id	short_name	keyword	url	date_created	d keywords.date_create
1	2 Google	google.com	{google:baseURL}...		0	NULL
2	7 princessauto.com	princessauto.com	https://...		13156209560872314	2017-11-26 22:39:20
3	8 Mozilla Support	support.mozilla.org	https://...		13069181148753723	2015-02-23 16:05:48
4	9 Living in Goma	livingingoma.com	https://...		13194672558136450	2019-02-15 02:49:18
5	10 metronews.ca	metronews.ca	http://...		13106406551637961	2016-04-29 12:29:11
6	11 New York Post	nypost.com	http://nypost.com/?...		13144522550824855	2017-07-14 16:15:50
7	12 GitHub	log2timeline.net	https://github.com/...		13132539422318790	2017-02-25 23:37:02

Fig. 8.25 "web data" SQLite file, "keywords" table

```
SELECT id, short_name, keyword, url,
       date_created,
       CASE date_created
             WHEN 0 THEN NULL
         ELSE datetime(date_created/1000000-11644473600,'unixepoch')
       END AS "Decoded keywords.date_created (UTC)"
FROM keywords
```

The above query produces Fig. 8.25.

There is value in querying this table alone. Note the created date in Fig. 8.25. One goes back to February of 2015. There most likely won't be any history entries for all the entries in this table, as history is only kept for 90 days and a user might not have navigated to some of the above in some time. But as you can see from the above output, this table retains the info indefinitely (until a user explicitly clears all their activity from their browser).

With the combined query we will look at next, it will only query the records in this table that have an associated entry in the "history" SQLite file in the "keyword_search_terms" table. Hence why querying the above separately can yield additional value.

Combining It Together to Show Intent

Querying across tables in different SQLite files is not a lot more difficult than simply querying across tables in a single SQLite file. If using something like DB Browser for SQLite, all you need to do is load the first SQLite file ("web data" in this case), and then attach the second one ("history" for this example) and give the second one a name.

In the following SQLite statement, "web data" SQLite is loaded, and then "history" SQLite is attached and given the name "history".

```
SELECT keywords.keyword AS "Search Engine",
       history.urls.url,
       history.keyword_search_terms.term,
       history.urls.visit_count,
       history.urls.last_visit_time,
        datetime(history.urls.last_visit_time/1000000-11644473600, 'unixe-
poch') AS "Decoded history.last_visit_time (UTC)"
FROM history.keyword_search_terms
       LEFT JOIN history.urls ON history.urls.id = history.keyword_search
_terms.url_id
       LEFT JOIN keywords ON history.keyword_search_terms.keyword_id
= keywords.id
```

The above yields the output in Fig. 8.26.

With this query, we can show intent. We see what the user searched for via the search box on various websites, how many times it was searched, and when it was last searched. If a particular result has a visit count of greater than 1 and you want to see all those records, you will have to re-run the above query and output the field "history.keyword_search_terms.url_id". Once you have the URL id, you can then query the "visits" table in the "history" SQLite file as we covered previously to see all the visits for that URL.

Missing Search Term

There will be situations when doing a search from within a search box on a site will not populate the search terms table. For example, during testing it was noted that if

	Search Engine	url	term	visit_count	last_visit_time	ɔded history.last_visit_time (
39	google.com	https://www.google.com/search?...	"chain of custody" gui...	3	13278261207008608	2021-10-09 13:53:27
40	google.com	https://www.google.com/search?...	"chain of custody" gui...	2	13278261214426295	2021-10-09 13:53:34
41	google.com	https://www.google.com/search?...	MS Word missing inser...	2	13278280963135611	2021-10-09 19:22:43
42	google.com	https://www.google.com/search?...	db browser	2	13278352786497213	2021-10-10 15:19:46
43	google.com	https://www.google.com/search?...	sqlite secure delete	2	13278353199026814	2021-10-10 15:26:39
44	google.com	https://www.google.com/search?...	sqlite secure delete	2	13278391570063453	2021-10-11 02:06:10
45	redirect.viglink.com	https://www.youtube.com/results?...	halloween makeup ...	1	13278465396549272	2021-10-11 22:36:36
46	youtube.com	https://www.youtube.com/results?...	halloween makeup ...	1	13278465396549272	2021-10-11 22:36:36
47	redirect.viglink.com	https://www.youtube.com/results?...	sqlite android tutorial	1	13278465434010116	2021-10-11 22:37:14
48	youtube.com	https://www.youtube.com/results?...	sqlite android tutorial	1	13278465434010116	2021-10-11 22:37:14
49	wikipedia.org	https://en.wikipedia.org/w/...	Garden State Parkway	1	13278465486249617	2021-10-11 22:38:06
50	wikipedia.org	https://en.wikipedia.org/w/...	Empire State Building	1	13278465554349494	2021-10-11 22:39:14
51	wikipedia.org	https://en.wikipedia.org/w/...	Empire State Building	1	13278465554349494	2021-10-11 22:39:14
52	yahoo.com	https://search.yahoo.com/search?...	Ford Mustang Mach E	1	13278465736465013	2021-10-11 22:42:16

Fig. 8.26 Querying across web data and history SQLite files

the search was done from the main Wikipedia page (Fig. 8.27), it did not populate the search term.

Searching via the search box on the above page (Fig. 8.27) did not populate the search terms table.

If the search was conducted from a language specific Wikipedia page, it did populate the search terms table. In the screenshot (Fig. 8.28), a search from the search box near the upper right corner would result in populating search terms.

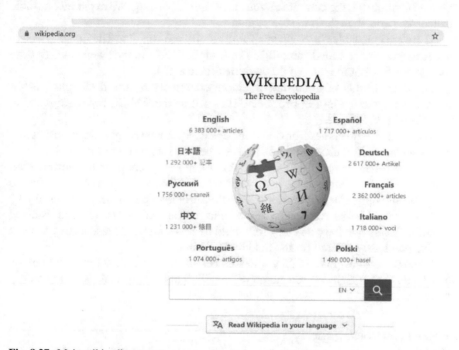

Fig. 8.27   Main wikipedia.org page

Fig. 8.28   en.wikipedia.org

8.9   Form Data

When you enter data in a field on a webpage, Google Chrome will save what you entered so that it can present it to you again in the future should you land on an input field by the same name. This is known as autofill, or form data.

Google Chrome tracks form data in an SQLite file called "web data" which we explored in part in the section about search terms. This SQLite file has 28 tables in Google Chrome 94, the current version at the time of writing. We explored one table called "keywords" in the previous section.

There are nine tables in "web data" with autofill information. The first one we are going to examine is called "autofill". The "autofill" table in the "web data" SQLite file has six fields at the time of this writing (Table 8.12).

The dates in this table are not the more commonly observed Google WebKit format. They are in Unix Epoch, but UTC same as the WebKit timestamps we've examined in this chapter.

The "name" field comes from the web page itself. A web programmer must assign a variable name to each input field. That is where this name comes from.

Note that there is no URL here. We don't know on what page this information was entered. All we know is the name of the field. When you visit a webpage for the first time and must enter information, Google Chrome might present options to you. Google Chrome looks at the name of the input field and checks for all records in the "autofills" table that have an entry for a field by this name and presents to you what you've previously entered for the field by that name.

In the screenshot (Fig. 8.29), we see the HTML source code for an input box on a website. On the right we see: name = "email". This tells us that the field name is "email".

Table 8.12 AUTOFILL table fields

Field	Description
name	The name of the field on the webform
value	The value that the user entered in that field
value_lower date_created date_last_used count	Same as "value", but all lower case The first date (Unix Epoch UTC) this autofill was used The last date (Unix Epoch UTC) this autofill was used The number of times it has been used

Fig. 8.29   autofill

Fig. 8.30 autofill for
"email" field name

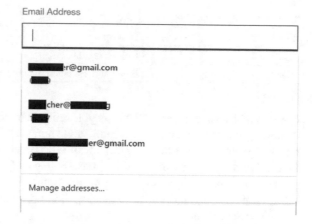

Google Chrome checks autofill for any entries where the name is "email" and presents those options to you in the pull down as observed in the screenshot (Fig. 8.30).

Well, the above is not accurate in this case. In this case Google Chrome pulled those values out of a different table, autofill_profile_emails. It's not clear to the author when Google Chrome uses the content of "autofill" versus "autofill_profile_emails".

We can just browse the content of "autofill" or use the following SQLite statement to decode the dates for us.

```
SELECT name, value, date_created,
DATETIME(date_created, 'unixepoch') AS "decoded date_created UTC",
date_last_used,
DATETIME(date_last_used, 'unixepoch') AS "decoded  date_last_used
UTC",
count
FROM autofill
```

In executing the above, the author saw that some of the records in the "autofill" table with the text 'email' for "name" had timestamps that decoded to as recently as September of 2021. So clearly Google is still using this "autofill" table in some cases. In the screenshot in this example (Fig. 8.30), there were email addresses in "autofill" that were not presented to the author as possible values. Instead, Google Chrome pulled the records from "autofill_profile_emails" to present previously entered values as options.

The "autofill_profile_emails" table in the "web data" SQLite file has two fields at the time of this writing (Table 8.13).

This table is one of five tables in "web data" relating to autofill: "autofill_profiles", "autofill_names", "autofill_phones", "autofill_addresses", "autofill_emails". Each

Table 8.13 AUTOFILL_PROFILE_EMAILS table fields

Field	Description
guid	Unique GUID
email	Email address entered by the user

of these tables have different fields as you can gather from the names of the tables. "autofill_phones" has two fields: "guid" and "number". "autofill_addresses" has 28 fields of which the first one is "guid". "autofill_names" has 19 fields, of which the first is "guid". And "autofill_profiles" has 18 fields, of which the first is "guid".

As you probably figured out by now, the "guid" is what links the records across these autofill tables. Across all these tables, there are 64 fields not counting the "guid" field in each of them.

The following SQLite command pulls some basic info from these five tables into one output.

```
SELECT use_count,
       origin,
       date_modified,
           datetime(date_modified, 'unixepoch', 'localtime') AS 'Decoded
date_modified (local time)',
       use_date, datetime(use_date, 'unixepoch', 'localtime') AS 'Decoded
use_date (local time)',
       autofill_profiles.guid,
       full_name,
       first_name,
       middle_name,
       last_name,
       street_address,
       city, state,
       zipcode,
       country_code,
       number, email
FROM autofill_profile_names
           JOIN autofill_profiles ON autofill_profiles.guid == autofill_
profile_names.guid
           JOIN autofill_profile_phones ON autofill_profiles.guid ==
autofill_profile_phones.guid
           JOIN autofill_profile_emails ON autofill_profiles.guid ==
autofill_profile_emails.guid
```

Those unfamiliar with SQLite might wonder why the table name does not precede the field names in the statement like we've seen in other SQLite statements. It's not necessary to put in the table name as the prefix if there is only one table with a field by that name. The risk in doing so here with so many fields is that if in the future a field is added to one of the tables we are querying that is the same as the field in another of the tables being queried, the query will fail as SQLite will not know from which table you want to query that field.

A more prudent practice when querying more than one table (which is almost always the case) is to use table_name.field_name rather than just field_name in your query.

The sample SQLite statement only queries 16 fields out of a possible 64. Thus, there is a lot more info in these autofill tables that you should examine if that type of information is potentially relevant to your case. These tables have fields with descriptive names, so you should have no difficulty simply browsing them without needing to use an SQLite statement.

8.10 Bookmarks

Bookmarks in Google Chrome are not stored in an SQLite as is the case in Firefox. Instead, they are stored in a JSON file. You can view it same as what we did for Local State and the Preferences file, by adding the extension ".json" to it and then dragging it into Firefox with the extension JSONVIEW, or using Notepad++ with the appropriate addon.

Within this JSON file you will see a series of key:value pairs, where a value can be another key:value pair, and that value can be yet another key:value pair, and so on.

The author is not aware of any native query language for JSON like we have for SQLite. There are solutions in various stages of development such as JSONPath and json-query. If you are a Python coder, you can use the JSON library to read a JSON file. The library parses it out to a Python dictionary with embedded dictionaries, lists, and values. From there, querying it becomes much more intuitive to navigate if you are familiar with Python dictionaries and lists.

Within the JSON structure for Google Chrome bookmarks a folder/sub-folder will have the following attributes:
"date_added": "13245036499010868",
"date_modified": "13277258678335801",
"guid": "60f9b262-f2dc-40c0-a7e7-18e668124622",
"id": "593",
"name": "Work",
"type": "folder"
The dates are Google WebKit UTC timestamps. The name is what you see in the GUI when viewing your bookmarks.

An individual bookmark entry has similar data but note that it only has one timestamp.

"date_added": "13259352719115332",
"guid": "a201dcd4-6b91-48e4-abb4-9513bccd5dfb",
"id": "711",
"name": "dates in hiding Archives—Metaspike",
"type": "url",
"url": "https://www.metaspike.com/tag/dates-in-hiding/"

It's possible to use Excel to import and parse a JSON file via the Get Data feature, much like you could use the free desktop version of PowerBI to accomplish something similar. But when briefly tested by the author, although all the bookmarks and folders were parsed out nicely, the hierarchy was lost. You could not determine from the Excel sheet which URL belonged to which folder.

Perhaps a more skilled Excel user might know how to retain that structure while importing a JSON file. Once you have it in Excel, you can add a column to convert a WebKit timestamp to a format that Excel can display in human readable format.

You can use the following Excel formula to convert a WebKit time to a Unix time.

=(CELL/1000000-11644473600)/60/60/24+"1 jan 1970" [26]

Where CELL is the cell you wish to convert.

Excel can convert the result to human readable via the format cell option. For example, you can format it as "YYYY-MMM-DD HH:mm:ss.000 UTC" to display it with the four digit year, the month abbreviation, the day, and the time in 24 h format with leading 0 and include milliseconds.

To preserve the folder structure of bookmarks, you'll need a tool that parses it, or write your own Python statement to parse it. The author doesn't use either, as the evidence is typically not found in bookmarks. If it is, it's usually a single entry that you can easily manually parse out. If you need to produce all the bookmarks, you'll want a tool to help you do that efficiently.

8.11   Other

This section will briefly touch on other browser artifacts, but will not look at how to analyse any of them as they are not database related, or more advanced.

Incognito Mode

Like all other modern browsers, Google Chrome supports browsing in private mode where nothing is saved to the drive. Google calls this Incognito Mode and they provide a detailed answer on what it does, and doesn't, do [27].

Incognito mode doesn't stop a network appliance from monitoring your traffic. It doesn't stop websites from collecting data about your browsing session. It also apparently doesn't stop (or at one time didn't stop) Google from engaging in analytics about your usage of their browser. Evidence of this from a lawsuit working its way through the courts [28].

Beyond the above, there are forensic tools that will look for evidence of Incognito browsing. Presumably it looks for this in the swap file, or unallocated space, as Google Chrome does not write anything to the user profile databases when browsing in this mode.

Cache

Google Chrome saves a local version of pages you browse in its local cache stored on your device so that if you visit it again before the cache expires, it can serve up the page from cache rather than from the source webpage. This was especially important in the early days of the Internet when it was slow, and you had a monthly bandwidth cap.

That's no longer a concern for many Internet users in first world countries. But there are still areas in both first world and third world countries where browser cache still serves its purpose.

Analysis of cache is something few people will ever attempt manually, the author included. Because of this, we will leave this topic for those who want to walk a path seldom traveled by their peers.

Session Recovery

When you browse with Google Chrome, it keeps a copy of your current browsing session in a session recovery file. It does this to be able to recover a browsing session from a crash, as well as to pick up where you left off if you configure Google Chrome's start up page to "Continue where you left off".

There is a 2012 post from Alex Caithness of CCL Forensics [29] where Alex takes us along on his journey to decode Google Chrome's session recovery. It's a deep dive into this artifact and in reading it, you will quickly conclude that it's not something you'll want to attempt to manually analyze, as it's only evolved since 2012. Within this file you can not only find your currently open web pages, but the tab history for each tab (up to 50 tab history entries per tab), form data, cookies, and probably other stuff.

If you hit on a keyword in one of Google Chrome's session recovery files, hopefully you have a tool that will parse it for you. There is a way you can examine at least the URLs in each tab without resorting to trying to decode it manually or with a specialized tool. You could export a Google Chrome profile from a forensic image and copy it over an existing Google Chrome profile and ensure you have the correct preferences to cause Google Chrome to "Continue where you left off", resulting in Google Chrome opening the session recovery file for you and opening all the tabs that were open, allow you to open closed tabs saved in the session recovery file, and even allow you to navigate through the tab history (forward/back) in each tab.

If a person had 10 tabs open and each had the maximum 50 entries for tab history, that's 500 URLs in that session recovery file. So, if you need to examine the session recovery file because it contains a keyword of interest to you, you could always use the shoestring option suggested herein in absence of a tool that can decode it for you.

Saved Passwords

Google Chrome saves passwords to an SQLite file, but as expected, they are stored in an encrypted format. You could examine the SQLite file and try and decrypt the passwords if tools are available to do that. Or you could actually use the similar approach to what the author suggested to examine the session recovery file.

You could copy out the Google Chrome profile from your image and into a profile on your forensic machine and then go into the browser settings to view saved passwords. If that doesn't work and your situation warrants it, you could navigate to the login page for the password you are interested in and let Chrome auto populate the username and password.

The password will be masked with "dots" as you know from experience. But there is a simple webpage hack you can do on a live page to reveal the password behind the dots. There are many articles and YouTube videos that show you how to do this, so we won't bother covering it here.

8.12   Summary

Google Chrome is a very popular browser. Knowing where it stores many of its artifacts and how to manually access them will help you be a better forensic examiner, even when using tools that do most of the heavy lifting for you. It will help you recognize if your tool is not parsing something correctly, or if it's only parsing some of the info that's available to you.

If you know how to analyst Google Chrome on one platform, you know how to analyze it on all platforms. And by knowing how to analyze Google Chrome artifacts, you will be well equipped to also analyze common artifacts in Microsoft Edge, Opera, Brave, and any other browser that is based on Chromium.

Acknowledgements The first author wrote a Google Chrome guide in November of 2012. This 29-page guide was used by the author when teaching browser forensics on the Internet Evidence Analysis Course at the Canadian Police College. Since then, he has updated the guide several times. The most recent update was released in December of 2019 thanks in large part to the research the author did as part of his dissertation for his M.Sc. in Forensic Computing and Cybercrime Investigations from the University College of Dublin. The guide has grown to 187 pages and continues to be used as a resource when teaching browser forensics at the Canadian Police College.

References

1. https://www.internetlivestats.com/
2. https://www.varonis.com/blog/cybersecurity-statistics/
3. https://www.statista.com/statistics/873097/malware-attacks-per-year-worldwide/

4. Europol. Europol identifies 3600 organised crime groups active in the EU. [Online]. Available https://www.europol.europa.eu/content/europol-identifies-3600-organised-crime-groups-active-eu-europol-report-warns-new-breed-crim

5. Oh, J., Lee, S., & Lee, S. (2011). Advanced evidence collection and analysis of web browser activity. *Digital Investigation, 8*, s62–s70.2. S. Canada. (2018). *Canada at a glance 2018*. [Online]. Retrieved March 19, 2022, from https://www150.statcan.gc.ca/n1/pub/12-581-x/201 8000/pop-eng.htm

6. Warren, C., El-Sheikh, E., & Le-Khac, N.-A. (2017). Privacy preserving internet browsers—Forensic analysis of Browzar. In K. Daimi, et al. (Eds.), *Computer and network security essentials* (pp. 369–388, 18 pages). Springer Berlin Heidelberg. https://doi.org/10.1007/978-3-319-58424-9_21

7. Reed, A., Scanlon, M., & Le-Khac, N. A. (2017). Forensic analysis of epic privacy browser on windows operating systems. In *16th European Conference on Cyber Warfare and Security*, Dublin, Ireland, June 2017.

8. https://www.chromium.org/

9. https://developer.mozilla.org/en-US/docs/Glossary/Blink

10. https://webkit.org/

11. https://www.magnetforensics.com/blog/artifact-profile-google-chrome/

12. https://www.json.org/

13. https://addons.mozilla.org/en-US/firefox/addon/jsonview/

14. https://www.epochconverter.com/

15. https://www.epochconverter.com/WebKit

16. https://www.sqlite.org/pragma.html

17. https://www.sqlite.org/pragma.html#pragma_secure_delete

18. https://www.sqlite.org/lang_vacuum.html

19. https://sqlitebrowser.org/

20. https://source.chromium.org/chromium/chromium/src/+/main:components/history/core/bro wser/url_row.h;l=92?q=hidden%20history&ss=chromium, lines 91–92, 144–146.

21. https://source.chromium.org/chromium/chromium/src/+/master:ui/base/page_transition_ types.h

22. https://source.chromium.org/chromium/chromium/src/+/master:components/history/core/bro wser/history_types.h, lines 46–51.

23. https://source.chromium.org/chromium/chromium/src/+/main:components/history/core/bro wser/download_database.cc?q=%22create%20table%20downloads%22&ss=chromium%2Fc hromium%2Fsrc&start=1, lines 314–345.

24. https://source.chromium.org/chromium/chromium/src/+/main:components/download/public/ common/download_interrupt_reason_values.h

25. https://source.chromium.org/chromium/chromium/src/+/main:components/sync/proto-col/ search_engine_specifics.proto?q=search_url_post_params&ss=chromium, lines 21–94.

26. https://timothycomeau.com/writing/chrome-history

27. https://support.google.com/chrome/answer/7440301

28. https://www.searchenginejournal.com/google-to-face-5b-lawsuit-over-tracking-users-in-inc ognito-mode/399113/

29. https://digitalinvestigation.wordpress.com/tag/chrome/

Printed in the United States
by Baker & Taylor Publisher Services